T0291714

Beyond Versus

Life and Mind: Philosophical Issues in Biology and Psychology
Kim Sterelny and Robert A. Wilson, Series Editors

Beyond Versus: The Struggle to Understand the Interaction of Nature and Nurture,
James Tabery, 2014

Investigating the Psychological World: Scientific Method in the Behavioral Sciences,
Brian D. Haig, 2014

Evolution in Four Dimensions: Genetic, Epigenetic, Behavioral, and Symbolic Variation in the History of Life, revised edition,
Eva Jablonka and Marion J. Lamb, 2014

Cooperation and its Evolution,
Kim Sterelny, Richard Joyce, Brett Calcott, and Ben Fraser, editors, 2013

Ingenious Genes: How Gene Regulation Networks Evolve to Control Development,
Roger Sansom, 2011

Yuck! The Nature and Moral Significance of Disgust,
Daniel Kelly, 2011

Laws, Mind, and Free Will,
Steven Horst, 2011

Perplexities of Consciousness,
Eric Switzgebel, 2011

Humanity's End: Why We Should Reject Radical Enhancement,
Nicholas Agar, 2010

Color Ontology and Color Science,
Jonathan Cohen and Mohan Matthen, editors, 2010

The Extended Mind,
Richard Menary, editor, 2010

The Native Mind and the Cultural Construction of Nature,
Scott Atran and Douglas Medin, 2008

Describing Inner Experience? Proponent Meets Skeptic,
Russell T. Hurlburt and Eric Schwitzgebel, 2007

Evolutionary Psychology as Maladapted Psychology,
Robert C. Richardson, 2007

The Evolution of Morality,
Richard Joyce, 2006

Evolution in Four Dimensions: Genetic, Epigenetic, Behavioral, and Symbolic Variation in the History of Life,
Eva Jablonka and Marion J. Lamb, 2005

Molecular Models of Life: Philosophical Papers on Molecular Biology,
Sahotra Sarkar, 2005

The Mind Incarnate,
Lawrence A. Shapiro, 2004

Organisms and Artifacts: Design in Nature and Elsewhere,
Tim Lewens, 2004

Seeing and Visualizing: It's Not What You Think,
Zenon W. Pylyshyn, 2003

Evolution and Learning: The Baldwin Effect Reconsidered,
Bruce H. Weber and David J. Depew, editors, 2003

The New Phrenology: The Limits of Localizing Cognitive Processes in the Brain,
William R. Uttal, 2001

Cycles of Contingency: Developmental Systems and Evolution,
Susan Oyama, Paul E. Griffiths, and Russell D. Gray, editors, 2001

Coherence in Thought and Action,
Paul Thagard, 2000

Beyond Versus

The Struggle to Understand the Interaction of Nature and Nurture

James Tabery

The MIT Press
Cambridge, Massachusetts
London, England

This book was set in StoneSerif Std and StoneSans Std 9/13 pt by Toppan Best-set Premedia Limited.

Library of Congress Cataloging-in-Publication Data is available.
ISBN: 978-0-262-02737-3 (hardcover)
ISBN: 978-0-262-54960-8 (paperback)

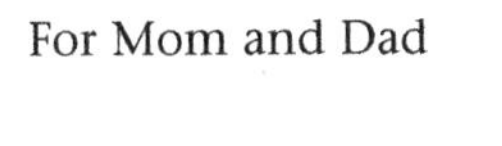
For Mom and Dad

Contents

Preface

Every book has a nature, and a nurture.

Its nature is that germinal idea, that nascent inclination to tell a story. The germinal idea for this book sprung from my fascination with the persistence of the nature/nurture debate. On the face of it, the nature/nurture debate is a scientific debate, and so it should be a debate about verified facts and empirical evidence. But if facts and evidence were all there is to the story, then two scientists (or two disciplines of scientists) should be able to look at the same data and reach the same conclusion. The persistence of the nature/nurture debate obviously suggests there is more to this story than just that. So what then is it? My goal was to expose the underlying philosophical disagreements that could lead two people (or two disciplines) to look at the same data about genes and the environment and reach very different conclusions. These disagreements involve disputes about what certain concepts mean, debates about how explanation works in science, arguments about what methodologies supply those explanations, and controversies about the ethical implications of the conclusions. My nascent inclination was to draw on the tools of the historian of science, the philosopher of science, and the bioethicist to disentangle these underlying disagreements so as to make room for disputants who were once talking past one another to meet on common ground.

A book's nurture includes all the people who nourished and shaped that germinal idea. Fortunately for me, this book was nurtured by others for close to a decade. A number of advisors, mentors, colleagues, friends, and family gave graciously of themselves to discuss these issues with me, read versions of this material, and provide me with invaluable feedback: Garland Allen, Kevin Amidon, Rebecca Anderson, André Ariew, Lisa Aspinwall, Margaret Battin, Teresa Blankenmeyer Burke, Jim Bogen, Mark Borrello, Jeffrey Botkin, Ingo Brigandt, Samuel Brown, Teneille Brown, Gretchen Case, Carl Craver, Thomas Cunningham, Lindley Darden, Stephen Downes, Linda Carr-Lee

Faix, Leslie Francis, Justin Garson, James Giordano, Paul Griffiths, Matthew Haber, Andrew Hamilton, Jonathan Hodge, Leslie Hogben, Eric Hutton, Annie Jamieson, Jonathan Kaplan, Brian Keeley, Maria Kronfeldner, James Lennox, Alan Love, Peter Machamer, Edouard Machery, Ron Mallon, Lucas Matthews, Erika Milam, Elijah Millgram, Sandra Mitchell, Lex Newman, John Norton, Robert Olby, Erik Parens, Lisa Parker, Laurence Perbal, Monika Piotrowska, Kathryn Plaisance, Anya Plutynski, Michael Pogue-Geile, Richard Purcell, Gregory Radick, Chris Renwick, Erin Rothwell, Susan Sample, Kenneth Schaffner, Thomas Schenkenberg, Jonah Schupbach, Jeffrey Schwartz, David Steffes, Jacob Stegenga, Karola Stotz, Omri Tal, Blake Vernon, C. Kenneth Waters, Mark Wicclair, Robert Wilson, and especially my wife—Dawn-Marie Tabery. I am especially grateful to Kenneth Schaffner at the University of Pittsburgh and Gregory Radick and Annie Jamieson at the University of Leeds for arranging to have students and colleagues go through an early version of the manuscript with a fine-toothed comb; I owe an additional debt to University of Utah students in my 2012 philosophy of science graduate seminar for doing the same. I also benefited from conversations or correspondence with a number of the scientists whose work is profiled in this book, including Avshalom Caspi, Roderick Cooper, Gilbert Gottlieb, K. Paige Harden, Kenneth Kendler, Terrie Moffitt, David Moore, Robert Plomin, Michael Rutter, and Daniel Weinberger. Thanks also to Pamela Speh for designing and creating figure 5.3. My research for this book was supported by the Tanner Humanities Center (University of Utah, 2013).

I am also grateful to the numerous publishers, archivists, and individuals who have given me permission to reproduce figures, correspondence, and previously unpublished material. Leslie Hogben gave me permission to quote from correspondence from Lancelot Hogben, archived at the University of Birmingham. The University of Adelaide gave me permission to quote from correspondence from R. A. Fisher, archived at the University of Adelaide. Cambridge University Press allowed me to reproduce figure 2.1, which was first published in Fisher and Mackenzie (1923), "Studies in Crop Variation, II," *The Journal of Agricultural Science* 13: 311–320. The American Philosophical Society Library granted me permission to quote from correspondence from Richard Lewontin and Theodosius Dobzhansky, archived at the American Philosophical Society Library. Elsevier gave me permission to reproduce figures in chapter 3, which were first published in Lewontin (1974), "Annotation: The Analysis of Variance and the Analysis of Causes," *American Journal of Human Genetics* 26: 400–411. Ken Kobre allowed me to reprint his wonderful photograph of Richard Lewontin for my figure 3.4. The American Association for the Advancement of Science permitted me to reproduce two figures for this book: figure 3.5 was first published in Layzer (1974), "Heritability Analyses of IQ Scores: Science or Numerology?" *Science* 183: 1259–1266, and figure 4.2 was first published in Caspi et al. (2003), "Influence of Life Stress on Depression: Moderation by a Polymorphism in the 5-HTT Gene," *Science* 301: 386-389. The University of Chicago Press permitted me to reproduce figure 5.1, which was originally published in Longino (2013), *Studying Human Behavior: How Scientists Investigate Aggression and Sexuality*. Chicago: University of Chicago Press. Oxford University Press permitted me to reproduce figure 5.2, which was first published in Craver (2007), *Explaining the Brain: Mechanisms and the Mosaic Unity of Neuroscience*. Oxford: Oxford University Press. The American Physiological Society allowed me to reproduce a portion of figure 5.3, which first appeared in Blum and Konnerth (2005), "Neurotrophin-Mediated Rapid Signaling in the Central Nervous System: Mechanisms and Functions," *Physiology* 20: 70–78. The Carnegie Institution of Science gave me permission to reproduce figure 6.1, which was first published in Clausen, Keck, and Hiesey (1940), *Experimental Studies on the Nature of Species, I. Effect of Varied Environments on Western North American Plants*. Washington, D.C.: Carnegie Institution of Washington. Wiley Periodicals, Inc. permitted me to reproduce figure 8.3, which was originally published in Bakermans-Kranenburg and van IJzendoorn (2006), "Gene-Environment Interaction of the Dopamine D4 Receptor (DRD4) and Observed Maternal Insensitivity Predicting Externalizing Behavior in Preschoolers," *Developmental Psychobiology* 48: 406–409. Finally, portions of three chapters in this book previously appeared in different forms. I am grateful to Springer for allowing me to incorporate material from my "Difference Mechanisms: Explaining Variation with Mechanisms" (*Biology & Philosophy*, 2009) into chapter 5, as well as from my "R. A. Fisher, Lancelot Hogben, and the Origin(*s*) of Genotype-Environment Interaction" (*Journal of the History of Biology*, 2008) into chapter 2. Oxford University Press allowed me to include material from my "From a Genetic Predisposition to an *Interactive Predisposition*: Rethinking the Ethical Implications of Screening for Gene-Environment Interactions" (*Journal of Medicine and Philosophy*, 2009) into chapter 7.

This book is a product of the interaction of its nature and its nurture—interaction between that germinal idea and all the people who have given of themselves to cultivating it. Of course, it's one thing to say that it is a product of the "interaction of nature and nurture"; it's another thing to understand what "the interaction of nature and nurture" actually means. I hope, in the pages to come, to make sense of just that.

1 Introduction

We have moved beyond *versus*. Whether it is medical traits like clinical depression, behavioral traits like criminality, or cognitive traits like intelligence, it is now widely recognized that "nature versus nurture" does not apply. There are no genes for depression such that having the gene ensures the development of depression and lacking the gene ensures resilience to depression. Likewise, there are no environments for depression such that all differences in depression can be explained by pointing to those differences in environment. Rather, it is a truism that these complex human traits arise from both nature and nurture, and differences in those traits arise from both differences in nature and differences in nurture. Since it is both, the challenge of explaining the development of a trait like depression or accounting for differences in depression must involve understanding the *interaction* of nature and nurture.

So what then is the interaction of nature and nurture?

Answering this question, it turns out, has not been easy. Indeed, it has created a great deal of controversy. For close to a century now, two radically different visions have struggled to define what the interaction of nature and nurture is, how it should be investigated, and what counts as evidence for it. This is a book about that struggle to understand the interaction of nature and nurture. It is a book about understanding why we have moved beyond versus and yet continue to find ourselves in the midst of the nature/nurture debate.

Francis Galton and the Origin of the Nature/Nurture Debate

To begin understanding the interaction of nature and nurture, it is best to take a step back to a time when nature and nurture really were seen as existing in competition. The phrase "nature versus nurture" is commonly attributed to Francis Galton (1822–1911), a younger cousin of Charles

Darwin. Darwin published his *On the Origin of Species* in 1859; it argued that, in nature, natural selection ensured the fit more often than the unfit passed their traits on to the next generation (Darwin 1859). Galton quickly became interested in how his cousin's theory of natural selection applied to human populations (Pearson 1914, 1924, 1930a, 1930b; Gillham 2001). If humans wanted to select for or against traits in their own population, Galton wondered, should that selection be focused on nature or on nurture?

Galton, however, was not the first individual to ask the question about nature versus nurture. Plato, in both the *Republic* and the *Phaedrus*, attributed to Socrates ruminations on the origins of traits such as courage (Colman and Woodhead 1989). And, in fact, even the nature/nurture alliterative phrase has precursors to Galton. In Shakespeare's *The Tempest*, Prospero described Caliban as "A devil, a born devil, upon whose nature/ Nurture can never stick" (Conley 1984; Teigen 1984).

These historical details do not detract from Galton's contribution to the history. For, even though earlier writers beat him to the question and to the phrase, Galton surely can be credited with first envisioning a science of nature versus nurture, a science that dealt with both understanding and intervention. With regard to understanding, Galton conceived of several methodologies designed to determine whether human traits ranging from intelligence to criminality were due to nature or nurture. In *Hereditary Genius* (1869), Galton collected information on several hundred years' worth of British judges, military commanders, scientists, and artists and concluded that these eminent examples were confined to a relatively small number of families, pointing to the "hereditary" origin of genius. The problem with Galton's family history approach to nature versus nurture was that individuals inherit both their nature and their nurture from their families, and so simply tracing genius down family trees did not disentangle whether it was nature or nurture that was responsible for the genius in each new generation. Recognizing this problem, Galton sought some new method that would allow him to measure nature and nurture. In the paper "The History of Twins," Galton recognized twins as nature's experiment and supplied case study after case study of twins' similarities (Galton 1875). As a result, Galton concluded, as he did in *Hereditary Genius*, that the evidence pointed in favor of nature being more responsible than nurture when it came to human traits (Burbridge 2001).[1]

Notably, Galton was not interested in the relationship between nature and nurture simply out of a curiosity for understanding the origins of human traits. Rather, it was intervention that first concerned him. Before "The History of Twins" and before *Hereditary Genius*, Galton advocated for actively shaping human populations to increase favorable traits and decrease unfavorable traits in the same manner that agricultural scientists selected for favorable traits and against unfavorable traits in crops and livestock (Galton 1865). The active shaping was to take place with what Galton later referred to as "eugenics" (Renwick 2011). Eugenics, meaning "good birth," was Galton's vision for a "science which deals with all influences that improve the inborn qualities of race" (Galton 1904, 1). By 1904, Galton's focus was on "inborn qualities" because, with the results of his family history and twin studies behind him, he thought efforts at increasing favorable traits and decreasing unfavorable traits had to be directed at nature rather than nurture. He thus advocated for programs "to bring as many influences as can reasonably be employed, to cause the useful classes in the community to contribute *more* than their proportion to the next generation" (ibid., 3). Such programs involved data collection (a "golden book" of thriving families), education (disseminating eugenics "like a new religion"), and intervention (socially condemning "unsuitable marriages" and praising "eugenic marriages"; ibid., 3–5).

Galton's vision for a science of nature versus nurture, his understanding of a winner in that battle, and his advocacy for a program that intervened on that understanding kicked off what we now think of as the modern nature/nurture debate, which persists into the present. Its first formulation, in the early decades of the twentieth century, pitted eugenicists against anti-eugenicists. Eugenicists and anti-eugenicists could agree that more intelligence and less criminality were good things. What they disagreed on was whether those traits arose from nature or nurture, and thus where the interventions should be directed. Eugenicists understood traits like intelligence or criminality and differences in those traits to result from nature and differences in nature, and so they encouraged interventions targeted there (e.g., more breeding from the intelligent upper classes, and less breeding from the criminal lower classes). Anti-eugenicists understood traits like intelligence or criminality and differences in those traits to result from nurture and differences in nurture, and so they encouraged interventions targeted there (e.g., more educational programs, and less discrimination).

From Versus to Interaction

It was in the context of the eugenics controversy that scientists first considered the possibility that traits and differences in those traits might arise, not from either nature or nurture, but from the interaction of nature and nurture. This book tells the story of the past, takes stock of the present, and considers the future of the interaction of nature and nurture. The book is organized into three parts: I start with a historical puzzle in part I, then draw on resources from the philosophy of science to resolve the puzzle in part II, and finally explore the ethical implications of getting that resolution right in part III.

The three chapters in part I investigate three separate episodes from the nature/nurture debate in which the interaction of nature and nurture figured prominently. In chapter 2, R. A. Fisher and Lancelot Hogben square off over heredity-environment interaction in the context of the eugenics controversy of the 1930s. In chapter 3, Arthur Jensen and Richard Lewontin debate genotype-environment interaction in the context of the IQ controversy during the 1970s. And in chapter 4, Terrie Moffitt and Avshalom Caspi, along with their critics and defenders, offer different visions of gene-environment interaction in the context of a debate over the causes of depression in the twenty-first century. Thus, there will be three different casts of disputants, three different contexts in which those disputants debate interaction, and three different eras in which those debates play out.

What is remarkable, then, is how similar these three debates are. As I will show, the same three questions arise again and again in each:

- the conceptual question—what is interaction?
- the investigative question—why and how should interaction be investigated?
- the evidential question—what is the empirical evidence for interaction?

And the same contradictory answers keep being offered over and over. On the one hand, one side of the interaction debates (consisting of Fisher, Jensen, and Moffitt and Caspi's critics) consistently conceptualized the interaction of nature and nurture as a purely statistical phenomenon that created an obstacle to getting on with the business of measuring the relative contributions of nature and nurture; on this side, interaction was treated as a potential nuisance but understood to be quite rare in nature. On the other hand, the other side (consisting of Hogben, Lewontin, and Moffitt and Caspi along with their defenders) consistently conceptualized the interaction of nature and nurture as a developmental phenomenon that shed important light on the causal mechanisms of the developmental process; on this side, interaction was treated as vitally important to understanding the relationship between nature and nurture and understood to be quite common in nature. If I've successfully conveyed the nature of the interaction debates in these three chapters, then the polarization will start to feel repetitive, even frustratingly so. But that's the point—the particular formulation of the polarization in the interaction debates has been frustratingly resilient to changes in time and place.

This pattern of polarization raises several questions: what's going on here? Why would scientists who share a common interest in studying the relationship between nature and nurture reach such contradictory answers about the interaction of nature and nurture over and over again? What, in short, explains this divide? My goal in the chapters of part I is to answer these questions.

I am not the first to cover this terrain. A number of explanations of this divide have been offered over the years by historians of science, philosophers of science, and even scientists who have participated in the interaction debates themselves. Some have appealed to different political motivations as the source of the divide; on this account, different answers arise because, for instance, one disputant is a capitalist while the other disputant is a Marxist. Others have appealed to a racist divide, where different answers arise because one disputant is a racist and the other is not. And still others have appealed to different intellectual aptitudes as the source of the divide; with an intellectual divide, different answers arise because, for example, one side knows how to design a meta-analysis while the other side does not.

I intend to explain the divide by appeal to something else. The disputants in the interaction debates have been separated by an *explanatory divide*—a disagreement concerning how explanation works in science. This disagreement stems from the fact that the disputants came to consider the interaction of nature and nurture from within two quite distinct approaches to studying the relationship between nature and nurture—what I will call the "variation-partitioning approach" and the "mechanism-elucidation approach." Variation-partitioners and mechanism-elucidators have both been around throughout the entire history of the nature/nurture debate, and they share a common interest in studying the relationship between nature and nurture. But they go about that study in two very different ways: they identify different things that need explaining; they ask different causal questions about the thing that needs explaining; they point to different things that do the explaining; and they utilize different methodologies to provide those explanations. An example will help convey this, so consider depression again. Members of both the variation-partitioning approach and the mechanism-elucidation approach are interested in studying the nature and nurture of depression, but that commonality belies important differences. On the variation-partitioning approach, the thing to be explained is variation in depression in some population; employers of this approach ask *how-much* questions about that variation—how much of the variation in depression is the result of differences in nature and how much is due to differences in nurture? The variation-partitioners answer those how-much questions by identifying and measuring the causes of variation responsible for the variation in depression, and they utilize statistical methodologies to generate that measurement by partitioning up the total variation in depression and allotting it to the assorted causes of variation. On the mechanism-elucidation approach, in contrast, the thing to be explained is the developmental process that gives rise to depression; employers of this approach ask *how* questions about that process—how do differences in nature and differences in nurture interact during the developmental process to give rise to differences in depression? The mechanism-elucidators answer those questions by elucidating the causal mechanisms responsible for depression, utilizing experimental methodologies that intervene on the mechanisms to generate that elucidation.

The explanatory divide I've described above shaped divergent understandings of the interaction of nature and nurture throughout the nature/nurture debate. On the variation-partitioning approach, the interaction of nature and nurture created an obstacle to getting on with the business of partitioning variation (the explanatory focus of that approach); as a result, it was treated as a nuisance by variation-partitioners and dismissed as a distraction. The variation-partitioners have long had their own concept of interaction, which I will call the "biometric concept of interaction." On the mechanism-elucidation approach, though, the interaction of nature and nurture was seen as a valuable insight into the developmental relationship between nature and nurture and sought out for the information it provided about the causal mechanisms of the developmental process (the explanatory focus of that approach). The mechanism-elucidators have likewise had their own concept of interaction quite distinct from the variation-partitioners, which I will call the "developmental concept of interaction." In chapters 2, 3, and 4, I will show how the disputants in the interaction debates have been largely talking past one another over the last century. They were talking past one another because, even though they were prima facie debating interaction, in actuality they were employing different concepts of interaction, they were offering different rationales for why they investigated interaction, and they were judging the empirical evidence of interaction differently.

My exposition of the explanatory divide in the chapters of part I helps us to understand why the same opposing answers to the conceptual, investigative, and evidential questions keep arising over and over again. But this explication does nothing to resolve that divide. The resolution comes in the chapters of part II, where the goal is to bridge the explanatory divide by way of integrating the variation-partitioning and mechanism-elucidation approaches along with their distinct concepts of interaction. A bridge is a means of spanning two otherwise isolated locations. The variation-partitioning approach, along with its biometric concept of interaction, and the mechanism-elucidation approach, along with its developmental concept of interaction, have been isolated for quite some time.

In chapter 5, I hold off on discussing interaction in favor of focusing on the variation-partitioning and the mechanism-elucidation approaches more generally. The interaction of nature and nurture has been one feature of the debate between these two approaches, but it is not the only feature. And so temporarily suspending discussion of interaction in chapter 5 allows me to take a step back and construct a more general bridge between these approaches to studying nature and nurture. The explanatory bridge I construct in chapter 5 is designed to explicate an integrative relationship between the two approaches' separate things to be explained (variation in a population versus developmental process), causal questions about those things (how much versus how), things that do the explaining (causes of variation versus causal mechanisms), and methodologies that provide the explanation (statistical versus interventionist). To achieve this integration, I draw on recent developments from the philosophy of science. In particular, I will utilize a philosophical move made by C. Kenneth Waters to characterize precisely what the variation-partitioning approach contributes to studying nature and nurture, and I will utilize research from the philosophy of mechanisms to characterize precisely what the mechanism-elucidation approach contributes to studying nature and nurture. Uniting the contributions of Waters with the philosophy of mechanisms provides the resource for integrating the partitioning of variation and the elucidation of mechanisms.

With the relationship between the variation-partitioning and mechanism-elucidation approaches developed in chapter 5, I return in chapter 6 to the conceptual, investigative, and evidential questions that were productive of so much controversy throughout the interaction debates. Members of the variation-partitioning approach and members of the mechanism-elucidation approach consistently conceptualized interaction differently, promoted different means of investigating it, and offered different assessments of its prevalence in nature. In contrast to the apparently incompatible answers offered by the disputants, I offer integrative answers to these questions that draw on the resources from both approaches. For example, members of the variation-partitioning approach repeatedly conceptualized the interaction of nature and nurture as an absence—the absence of the separate contributions of nature and nurture; members of the mechanism-elucidation approach, in contrast, repeatedly conceptualized the interaction of nature and nurture as a presence—the presence of a uniquely developmental source of variation. The concept of interaction that I provide in chapter 6 captures why both sides were getting something right and something wrong—because interaction is both a presence (of what I will call "interdependent actual difference makers") and an absence (of "independent actual difference makers"). Likewise, members of the variation-partitioning approach routinely pointed to a scientific result where no interaction turned up and claimed that result showed interaction was rare in nature; members of the mechanism-elucidation approach instead eagerly pointed to a scientific result where interaction did turn up and claimed that result showed interaction was common in nature. My review of nine decades' worth of empirical evidence on the interaction of nature and nurture points to a much more nuanced conclusion. Whether it was research on crop yield in plants, meat production in livestock, or behavioral traits in humans, there has been plenty of empirical evidence for interaction, plenty of empirical evidence against interaction, and no warrant for assuming ahead of time whether any particular trait will or will not result from some particular interaction of nature and nurture.

In the final two chapters, I shift attention to what the future holds for interaction, particularly concerning struggles over what ethical issues will be raised by new discoveries of interaction between particular genes and particular environments. Bioethicists interested in the interaction of nature and nurture have been drawn to high-profile cases of interaction involving controversial traits with controversial interventions. For example, a case of interaction between a particular gene (*MAOA*) and a particular environment (childhood maltreatment) was linked to antisocial behaviors such as criminal violence in 2002; individuals with one version of the gene who were maltreated as children were much more likely to engage in criminal violence later in life than individuals with the other version of the gene (Caspi et al. 2002). In response, bioethicists wondered whether or not prospective parents would screen embryos during preimplantation genetic diagnosis to avoid giving birth to a child with the "genetic predisposition to violence," and others wondered whether or not states would screen newborns to identify those children who carry the "genetic predisposition to violence." Several bioethicists even advocated for these measures. I will show in chapter 7 why the future does not hold in store such interventions; it does not hold preimplantation genetic diagnosis or state newborn screening for a genetic predisposition to violence because no such genetic predisposition to violence has ever been identified. The very language of a "genetic predisposition to violence" mischaracterizes the results of the 2002 study that received so much attention. The purpose of chapter 7 is thus to explicate the dangers associated with mischaracterizing the interaction of nature and nurture and to provide a conceptual tool for characterizing interaction correctly.

This is not to say that, once characterized correctly, the ethical issues raised by cases of interaction go away. But to find them, we must switch our attention from high-profile traits like criminality to less sensationalized traits like childhood allergies and temper tantrums, for allergies and oppositional behavior have both been linked to particular gene-environment combinations where parental decisions about the early experiences and exposures of their children shape the nurture side of these gene-environment interactions—parental decisions such as whether to add a dog or a cat to the family, whether to send an infant to daycare, and whether to respond to a toddler's temper tantrum with empathy or punishment. Depending on which version of various genes a child may have, parental decisions about these childhood experiences could increase, decrease, or have no effect at all on the risk of allergies and oppositional behavior. So the question is: might parents want to obtain this genetic information about their child so as to guide these decisions? I will show in chapter 8 that, unlike statewide newborn screening for a genetic predisposition to violence, the potential for a "genetic guide to parenting" is a very real possibility that deserves attention from bioethicists.

Scope of This Book

Before proceeding to part I, it will be helpful for me to make clear the scope of this book. It is a book about the interaction of nature and nurture, but there is much that could conceivably fall under that umbrella which I am not able to address for reasons of space and maintaining a coherent narrative. For example, disputes surrounding interaction are not confined to the nature/nurture debate. Epidemiologists, especially in the field of public health, have had extensive discussions about the nature of interaction.[2] The scope of this book is not interaction between any two variables; it is interaction between two specific variables, nature and nurture, and so I do not engage those discussions of interaction in other domains except where they intersect with and affect the nature/nurture debate. Likewise, the book is not meant to be a complete history of research on the interaction of nature and nurture; I focus specifically on controversial episodes from the nature/nurture debate in which the interaction of nature and nurture figured prominently. As a result, research that did not take place during these controversial episodes has been left out.

I will also not be able to consider evolutionary explanations of interaction in this book. The concept of phenotypic plasticity is closely associated with the concept of interaction. Phenotypic plasticity refers to cases where a gene or genotype responds differently (or "plastically") to different environments. There has been quite a bit of work done on the evolutionary implications of phenotypic plasticity, in particular on the adaptive advantages of phenotypic plasticity (Pigliucci 2001; DeWitt and Scheiner 2004). The nature/nurture debate is about the causes of traits and differences in those traits in current populations, and the interaction debates are about the interaction of nature and nurture in those current populations. The discussions of phenotypic plasticity, though fascinating, are more about how a population's plastic response to the environment shapes the evolution of that population. And so I will not be able to discuss that work here either.

Finally, I want to be clear at the outset that this book is no attempt to "solve" the nature/nurture debate. When I told people that I was writing a book about the nature/nurture debate, I commonly got asked, "Who wins? Nature or nurture?" With the interaction of nature and nurture, though, there is no traditional winner, since the effects of nature and the effects of nurture are interdependent. Some commentators on the nature/nurture debate have argued that everyone now agrees with this "interactionist credo," simultaneously wondering why the debate persists despite this consensus (Kitcher 2001). What I hope to show in the early chapters of part I is that the debate persists precisely because "interaction" has meant different things to different people. So when I talk about "resolving" the interaction debates in part II, what I hope to do is clear up that confusion by drawing on the tools of history and philosophy, and explore the ethical implications of that resolution in part III with the tools of bioethics.

Reading This Book

The three parts of this book roughly break down into history of science (part I), philosophy of science (part II), and bioethics (part III). As a result, I hope readers from a number of different backgrounds will be interested in the story—scholars from across the humanities, scientists who perform research on nature and nurture, and non-academic readers simply interested in the frustrating persistence of the nature/nurture debate. The book is designed to be read from beginning to end—from historical puzzle in part I, through philosophical resolution in part II, to bioethical implications in part III. But I also understand that some readers will be interested in only certain parts (just the history, or just the contemporary bioethical implications, for example). To accommodate those readers, I have included part introductions that briefly summarize the material from the previous chapters and motivate the transition to subsequent chapters (the exception is part I, where this introductory chapter is meant to provide that resource). The part introductions may feel redundant for those of you who are reading the book from beginning to end, but I hope that this redundancy can be forgiven in light of my goal of serving different readers with diverse interests.

I The Interaction Debates

2 The Origin(*s*) of Interaction: Interaction in the Eugenics Controversy

In the early decades of the twentieth century, the eugenics movement swept across the United States and the United Kingdom.[1] Eugenicists worried that the "unfit" were outbreeding the "fit." In nature, natural selection ensured that the fit passed their traits on to the next generation more often than the unfit. But in human populations, eugenicists saw something different happening. In Britain, for eugenicists like Karl Pearson, who became the first Francis Galton Chair of Eugenics at University College London in 1911, the focus was on class; the concern was that governmental and charitable support for poor and "feebleminded" persons in the lower classes incentivized them to have more and more children, while cultural pressures on professional and cultural elites in the upper classes incentivized them to have fewer and fewer children (Porter 2004).[2] In the United States, for eugenicists like Charles Davenport, the focus was on race and ethnicity; the concern was that educated, upstanding whites were being out-bred by blacks and immigrants (Kevles 1995). What was the eugenicist to do?

Eugenicists on both sides of the Atlantic took a number of steps to spread the eugenic gospel, establishing institutes of eugenic research, eugenic societies, and eugenic journals. In Britain, Pearson set up the Eugenics Laboratory at University College London in order to collect human data and apply statistics to eugenic concerns (Porter 2004). In the United States, Davenport established the Eugenics Record Office at Cold Spring Harbor in Long Island, New York, to similarly collect data on the inheritance of wanted and unwanted traits (Allen 1986). In Britain, eugenicists set up the Eugenics Education Society with its associated *Eugenics Review* journal (Mazumdar 1992). In the United States, eugenicists founded the American Eugenics Society, which had its own *Eugenical News* (Osborn 1974). The gospel spread. Eugenic themes worked their way into everything from films, with titles like "Are You Fit to Marry?," to Sunday morning sermons, with society-sponsored eugenic sermon contests (Pernick 1996; Rosen 2004).

These efforts at educating the public, however, did not satisfy the eugenicists. More needed to be done. And so eugenicists began working closely with members of government to pass and enforce eugenically inspired laws. In Britain, members of the Eugenics Society helped pass the U.K. Mental Deficiency Act of 1913, which allowed for the forced institutionalization of feebleminded persons; in the 1930s, they fought to pass an act encouraging the voluntary sterilization of the feebleminded (Mazumdar 1992). In the United States, eugenicists helped pass the Immigration Restriction Act of 1924, which severely curtailed the influx of immigrants; in 1927, they worked with Virginia officials to defend the state's policy of involuntary sterilization of the feebleminded before the U.S. Supreme Court and won *Buck v. Bell*, a decision that sanctioned sterilization across the United States (Kevles 1995).

Ronald Aylmer Fisher (1890–1962) was a giant of twentieth-century science. He was the first to demonstrate a mathematical relationship between Darwinian natural selection and Mendelian genetics. He also created many of the statistical methodologies that continue to be taught in statistics classes around the world.

Fisher was also an ardent eugenicist. He was a member of the British Eugenics Society and even helped establish the Society's Committee for Legalising Sterilisation (Mazumdar 1992, 204). His statistical innovations were developed in part to make eugenic assessments of the relative importance of nature and nurture when it came to evaluating traits like feeblemindedness. In the process of developing those statistical innovations, Fisher quickly realized that the interaction of nature and nurture (or what he called the "deviations from summation formula") posed a potential problem for his methodologies; however, it was a problem he felt confident dismissing.

Lancelot Thomas Hogben (1895–1975), though not so well known as Fisher, made equally lasting contributions to science. A trained statistician and biologist, Hogben developed the African clawed frog (*Xenopus laevis*) as a model organism for both experimental research and pregnancy testing; *Xenopus* remains one of the most commonly used model organisms in biology (Gurdon and Hopwood 2000; Olszynko-Gryn 2013). He also founded the Society for Experimental Biology, which still exists today (Erlingsson 2009).

Hogben was also a fervent anti-eugenicist. Throughout the 1930s, Hogben led a vicious assault on the British eugenics movement generally and Fisher specifically. As Hogben saw it, the interaction of nature and nurture (or the "interdependence of nature and nurture," as he called it) was indeed a problem for Fisher's statistics and posed a serious challenge to the eugenicist.

A debate between Fisher and Hogben emerged, first in private communication and then in public, over this interaction of nature and nurture. What was interaction? Why and how should interaction be investigated? And what was the empirical evidence for it? Fisher and Hogben had radically different answers to these three questions. Where Fisher saw a potential but unproven nuisance, Hogben saw a fundamentally devastating critique. What explains Fisher and Hogben's different assessments of interaction? Was it simply that Fisher and Hogben were influenced by different religious or political motivations? A religious/political divide was surely part of this story, but only part. Fisher and Hogben, I argue, also faced an explanatory divide. I will show this by examining their separate routes to first recognizing the interaction of nature and nurture as a phenomenon. These routes were quite distinct and, as a result, when they finally came to discuss interaction, they understood the phenomenon quite differently: different concepts, different approaches to investigating it, and different verdicts on its prevalence in nature. This chapter about the origin of interaction is, thus, really a story about *origins*: Fisher's origin, and Hogben's origin.

The debate between Fisher and Hogben had a lasting impact on them personally and on the nature/nurture debate generally. On a personal level, the debate changed their relationship forever. Before the debate, they were cordial; afterward, they detested one another. On a wider level, though, Fisher and Hogben's separate routes to interaction created two very different ways of understanding the phenomenon that have persisted all the way to the present and shaped subsequent debates. The legacies of Fisher's route and Hogben's route, as well as the debates that ensued, will be the subject of the next two chapters. To understand those subsequent debates, though, we must start with the two scientists who laid the groundwork for debating interaction.

R. A. Fisher and the "Deviations from Summation Formula"

Fisher showed signs of mathematical genius early in life. As Joan Fisher Box, his daughter and biographer, tells it, he was interested in fractions by age three, considered astronomical calculations by age six, and enrolled in classes with boys years older than him by age ten (Box 1978, 12–13).[3]

Fisher's early achievement was especially remarkable given the fact that he suffered from extremely poor eyesight. "He began to wear spectacles of increasing power until the lenses were so thick that they resembled transparent pebbles," Box writes (ibid., 14). Because of his poor vision, Fisher developed an unorthodox approach to solving mathematical problems. Rather than writing out a mathematical proof in standard sequential steps, Fisher often worked the problem through in his head visually and then simply stated the answer. This was a source of some tension between Fisher and his teachers early in life; later in his life, it occasionally left his students and readers puzzling to follow his innovative reasoning (ibid.).

Despite his unconventional approach to problem solving, Fisher thrived at mathematics and matriculated at Gonville and Caius College, Cambridge, in 1909 on scholarship. He continued his mathematical studies there, adding physics toward the end of his stay (Hodge 1992). Fisher also became an enthusiastic eugenicist. He helped create the Cambridge University Eugenics Society in 1911, hosting meetings in his rooms, organizing public lectures by well-known eugenicists, assisting at the First International Eugenics Congress, and even delivering his own eugenic lectures (Mazumdar 1992, 97–105). After graduating in 1913, Fisher was eager to join England's military, which was preparing for World War I. Poor eyesight, however, disqualified him (Yates and Mather 1963, 92). Instead, he spent the war years performing a variety of jobs around London related to mathematics and eugenics. He also worked for the Eugenics Education Society, contributing to the society's eugenics bibliography and reviewing books for its *Eugenics Review* (Mazumdar 1992, 105).

"Analyzing the Causes of Variability"

In October 1918, at only 28 years of age, Fisher published "The Correlation between Relatives on the Supposition of Mendelian Inheritance" (Fisher 1918). At the time, Darwin's theory of natural selection and Mendel's principles of genetics were thought to be incompatible; the former treated variation in nature as gradual and continuous, while the latter treated variation as sporadic and discontinuous. A heated dispute emerged between biometricians, like Pearson, who backed Darwinian continuous variation, and Mendelians, like William Bateson, who backed Mendelian discontinuous variation.[4] Fisher's project was the resolution of this perceived incompatibility. It was an idea he first presented to the Cambridge University Eugenics Society in 1911.[5] Far from being incompatible, Fisher showed that the biometricians' statistical data on continuous variation could be derived from Mendelian inheritance (Norton 1978; Provine 2001).[6] Fisher's formulation of a mathematical relationship between Darwinian evolution and Mendelian genetics was the first in a series of contributions to what Julian Huxley later termed the "modern evolutionary synthesis" (Huxley 1942). He was soon joined in this work by the British polymath J. B. S. Haldane and the American geneticist Sewall Wright, who collectively created the field of population genetics (Provine 2001). This achievement alone would have secured Fisher's place in any history of twentieth-century science.[7]

Remarkably, synthesizing Darwinian evolution and Mendelian genetics was not the only feat accomplished in Fisher's 1918 article. In the process of deriving the mathematical relationship between Darwin and Mendel, Fisher also introduced a new statistical concept (Moran and Smith 1966; Kempthorne 1974; Box 1978). Fisher was interested in accounting for the sources of variation in a population. Traditionally, populations were statistically evaluated solely with an eye toward averages, but averages shed no light on variation. Fisher found that if a trait under investigation, such as height in humans, manifested itself in a population with a normal distribution (i.e., a bell curve), then the mean could be calculated along with the standard deviation. Fisher's novel contribution to the statistical analysis of variation in a population was to go beyond the standard deviation and analyze the square of the standard deviation. "When there are two independent causes of variability capable of producing in an otherwise uniform population distributions with standard deviations σ_1 and σ_2," Fisher wrote, "it is found that the distribution, when both causes act together, has a standard deviation $\sqrt{(\sigma_1^2 + \sigma_2^2)}$. It is therefore desirable in analyzing the causes of variability to deal with the square of the standard deviation as the measure of variability" (Fisher 1918, 399). Fisher, having introduced a new statistical measure, knew he had to simultaneously introduce a new statistical concept to capture this measure, and so he continued, "We shall term this quantity the Variance of the normal population to which it refers, and we may now ascribe to the constituent causes fractions or percentages of the total variance which they together produce" (ibid.). *Variance* was now available as a measurement for the statistician and the eugenicist.

The earlier generation of biometricians had already introduced the concept of the correlation coefficient as a numerical measure of association (Norton 1975; MacKenzie 1981b). Thus, correlation tables were, by 1918, common; and parental correlations (associations between parent and offspring) along with fraternal correlations (associations between siblings) were frequently calculated from these correlation tables by the biometricians. Fisher utilized this data for partitioning sources of variation in 1918 as a means toward assessing the relative importance of hereditary and environmental causes of variation when it came to height in humans, explaining, "For stature the coefficient of correlation between brothers is about .54, which we may interpret by saying that 54 per cent. of their variance is accounted for by ancestry alone, and that 46 per cent. must have some other explanation" (Fisher 1918, 400).

To what cause should this remaining 46% be attributed? Perhaps an environmental cause of variation? No! Fisher, the eugenicist, was quick to eliminate that possibility from the mind of his reader. "It is not sufficient to ascribe this last residue to the effects of environment. Numerous investigations by Galton and Pearson have shown that all measurable environment has much less effect on such measurements as stature" (ibid.). So with environmental variation expunged from the list of possible causes of variation, Fisher had to find another explanation for the 46% of the total variance left unaccounted. That other explanation came from Mendel's principles of segregation and dominance. Fisher calculated the variance between siblings attributable to Mendelian segregation and the effects of dominance. With variances due to ancestry, segregation ($1/2\ \tau^2$), and dominance ($3/4\ \varepsilon^2$) all accounted for, Fisher could finally sum up the sources of the total variance (ibid., 424):

Ancestry		54 per cent
Variance of sibship		
$\frac{1}{2}\ \tau^2$	31 per cent	
$\frac{3}{4}\ \varepsilon^2$	15 "	
Other causes	—	
		46 "
		100 per cent

Fisher, referencing the negligible variance left for "other causes," concluded, "it is very unlikely that so much as 5 per cent. of the total variance is due to causes not heritable, especially as every irregularity of inheritance would, in the above analysis, appear as such a cause" (ibid.). For Fisher in 1918, it was a clear win for the nature side of the nature/nurture debate.

As mentioned above, commentators have pointed to Fisher's 1918 paper for its dual achievement of (a) synthesizing Darwinian evolution and Mendelian genetics, and (b) introducing the statistical concept of variance. Less noticed, though crucial for the discussion of this book, is an additional phrase Fisher introduced to science in 1918—"cause of variability." Fisher wasn't just measuring variation; he was measuring *causes* of variation. The phrase pervaded his paper, as the quotes above attest: "when there are two independent causes of variability ... "; "it is therefore desirable in analyzing the causes of variability ... "; "it is very unlikely that so much as 5 per cent. of the total variance is due to causes not heritable." Fisher called his calculation above an "analysis of variance"; the great achievement, as he saw it, was not just that he introduced a measure of variation, but rather that he introduced a method for partitioning the *causes* of variation. This attention to identifying and partitioning causes of variability remained a constant throughout Fisher's career.

"A Great Complication"

An assessment of Fisher's evaluation of the relationship between hereditary and environmental causes of variation would be incomplete if it terminated with his conclusions made in 1918. Historians of genetics and eugenics have often characterized Fisher as a "reformed" eugenicist, emphasizing his ultimate recognition of the potential importance of environmental causes of variation (Soloway 1990; Barkan 1991; Mazumdar 1992; Kevles 1995). Pauline Mazumdar, in particular, has carefully detailed the evolution in Fisher's thought (Mazumdar 1992). According to Mazumdar, Fisher's 1918 paper was, from the very beginning, designed to accommodate the ideals of the Eugenics Society: (a) the compatibility of Darwin and Mendel, and (b) the negligible importance of environmental causes of variation (ibid., 110). But in 1919, Fisher left Cambridge to join the Rothamsted Agricultural Research Station in Harpenden (Box 1978; MacKenzie 1981b; Mazumdar 1992). At Rothamsted, Fisher's job was to examine environmental causes of variation rather than assume them to be a randomly distributed variable, as he had in his 1918 publication. So it was at Rothamsted that Fisher created many of the statistical methodologies, such as tests of significance and the design of experiments that continue to be used by statisticians today (Yates 1964; Box 1978; Johnstone 1987; Preece 1990). And it was there that Fisher first came to consider the problem posed by interacting causes of variation.

In 1918, Fisher explained that sources of variation could be summed as long as the causes of variability were independent. Prior to undertaking his work at Rothamsted, the environment could be treated as independent for the simple reason that Fisher took it to be negligible. In making no contribution to variability, there was no need for Fisher to concern himself with how environmental variation might be related to the other causes of variation. But at Rothamsted, with nurture now on the list of possible sources of variation, Fisher had to also consider the relationship between environmental variation and hereditary variation. He judged this possible complication in 1923 with Winifred A. Mackenzie, where they studied different potato varieties' responses to different manure-based fertilizer treatments (Fisher and Mackenzie 1923).[8] Fisher began by warning, "if important differences exist in the manurial response of varieties a great complication is introduced into both variety and manurial tests; and the practical application of the results of past tests becomes attended with considerable hazard" (ibid., 311). The "great complication" that posed a "considerable hazard" was the interaction of nature and nurture, with potato variety representing nature and fertilizer representing nurture in this case. If certain potato varieties always performed better than other potato varieties, or if certain fertilizers always performed better than other fertilizers, then the scientist (and the farmer) had a very easy decision to make when it came time to select which potato variety to plant and in which fertilizer. But if different potato varieties performed differently in different fertilizers, then there wasn't necessarily a "best potato" or a "best fertilizer"; the "best potato" depended on which fertilizer was used, and the "best fertilizer" depended on which variety of potato was planted. "Only if such differences are non-existent, or quite unimportant," Fisher cautioned, "can variety tests conducted with a single manurial treatment give conclusive evidence as to the relative value of different varieties, or manurial tests conducted with a single variety give conclusive evidence as to the relative value of different manures" (ibid.).

To test for this great complication, Fisher devised a method to evaluate the fertilizer responses of 12 different potato varieties. He divided a relatively small field (0.162 acres) into two equal parts, one part receiving a manurial treatment, the other receiving no treatment. Each half was then itself divided into 36 plots, and each of the 12 potato varieties then planted in triplicate within each field. Finally, each individual plot was divided again, so that three rows of seven plants were set in each plot; one row received only the basal manuring of the series to which it belonged, while the other two rows received in addition either potassium sulfate (muriate of sulfate) or potassium chloride (muriate of potash). With this experimental design, Fisher measured the weight of produce lifted from each of the rows, determining both the mean yield of each of the 12 varieties irrespective of the fertilizer applied (the "main effect" of variety), and the mean yield of each of the fertilizer treatments irrespective of the variety grown (the "main effect" of fertilizer). What followed was the first presentation

Variation due to	Degrees of freedom	Sum of squares	Mean square	Standard deviation
Manuring	5	6,158	1231·6	35·09
Variety	11	2,843	258·5	16·07
Deviations from summation formula	55	981	17·84	4·22
Variation between parallel plots ...	141	1,758	12·47	3·53
Total	212	11,740	—	—

Figure 2.1
Fisher's analysis of variance due to fertilizer manuring, potato variety, deviations from summation formula, and variation between parallel plots (reproduced from Fisher and Mackenzie 1923, table III).

of an analysis of variance table, listing the various causes of variation along with their respective contribution to the total variation in crop yield (see figure 2.1; Box 1978, 111).

The "deviations from summation formula" category was the measure of the differences between the potato varieties in their fertilizer response—the measure, that is, of interaction. It was called a "deviation" because Fisher's analysis of variance was designed to attribute variation first to the main effects (in this case, potato variety and fertilizer). If the variation due to potato variety and the variation due to fertilizer summed up to account for the total variation in crop yield, then there would simply be no variation left to attribute elsewhere, and so the result was referred to as "additive" because the main effects added up. Interaction arose in the analysis of variance if there was a deviation from the main effects adding up to fully accounting for the variation, in which case there was "non-additivity." So the question for Fisher was whether or not there was a deviation from this additive summation. In yet another innovative leap in this same article, Fisher determined that the deviations from the summation formula were not significantly greater than would occur by chance, leading him to conclude, "In the present material evidently the varieties show no difference in their reaction to different manurial conditions" (Fisher and Mackenzie 1923, 317). This comparison of a source of variation against chance was an early test of statistical significance, or what is now called an "F-test" in honor of Fisher's development of the method. For Fisher, in 1923, it justified dismissing the problem.

Fisher paid little attention to the great complication posed by interacting causes of variation for the next ten years; the considerable hazard had been examined and dismissed. In *Statistical Methods for Research Workers*, his extremely influential how-to guide for statistics, Fisher made only passing reference to the "interaction of causes" just to point to the results of his 1923 publication with Mackenzie and then dismiss the phenomenon (Fisher 1925, 209). And in his eugenic writings from this period on the sterilization of the feebleminded and the high fertility of people in lower classes, Fisher made no reference to interaction (Fisher 1924a, 1924b, 1926a, 1926b). He was not concerned that people in a lower class might become people of a higher class in a different environment, or that the feebleminded might become less feeble in a different environment. It took a letter from Hogben, a decade later, for Fisher to reconsider the issue.

Lancelot Hogben and the "Interdependence of Nature and Nurture"

Hogben described his "larval existence" like that of many prominent biologists: obsessively collecting beetles and butterflies, identifying birds and recognizing them by their eggs, and exploring local geography. "I wanted to be a biologist long before I was twelve," he recalled 60 years later (Hogben 1998, 2).[9] Biology, however, was not what God intended for Hogben ... at least that was how his mother saw it. He was born two months prematurely, and to ensure that he would survive the ordeal, his mother dedicated him from birth to becoming a missionary (ibid., 1). This religious devotion was no less powerful on the paternal side of his parenting. Thomas Hogben,[10] a self-employed Methodist preacher, spent his days ministering to seamen at the local port under a banner extolling the benefits and burdens of the Christian God. "In the foreground was the lake of brimstone and fire. Across the middle was the edge of a cliff where stood the theatre, the brothel, the casino, the racecourse, the tavern, the *Palais de Danse* and other haunts of Satan. From the edge of the cliff the lost departed were falling in different stages of incandescence. Above the cliff was a solitary pilgrim pursuing a winding road to the rising sun; and, ironically, below it across the flames the legend: *God is Love*" (ibid., 4). Fortunately, the young Hogben and his parents were able to reach a compromise during these formative years; the field of medicine allotted the boy the time to study biology while also preparing him for service as a medical missionary (ibid., 13).

Largely self-educated at the Stoke Newington Public Library in northeast London, Hogben excelled academically and won a scholarship to attend Trinity College, Cambridge, in 1913, just as Fisher was graduating (ibid., 24–25). At Cambridge, Hogben cultivated his biological interests and replaced his parents' religious zeal with a devotion to socialism. He studied botany, physiology, zoology, and embryology (ibid., 40–41). And he entered social life with an equal vigor. Assessing the social societies available to him at the time, Hogben recalled, "I still regard the Union Debating Society of

Cambridge (even more that of Oxford) as a potting shed for the cultivation of mentally retarded politicians. The most lively discussions at an intellectually high level were those which took place at the *Moral Sciences*, colloquially *Moral Stinks, Club*, where Bertrand Russell and [G. E.] Moore minced words with their philosophical competitors, in the Fabian Society and its study circles, and in the *Heretics* founded by C. K. Ogden of Basic English fame" (ibid., 33). The Fabian Society was a particularly accommodating match for Hogben; he met his first wife, Enid Charles, there and eventually became its secretary, changing the society's name to the University Socialist Society (ibid., 51).

Unlike Fisher, the pacifist Hogben actively avoided military combat. At the outset of World War I, Hogben joined noncombatant Quaker relief organizations. When the British government introduced compulsory military service in 1916, Hogben protested this action as a conscientious objector and spent several months imprisoned in London's Wormwood Scrubs for the decision (ibid., chapter 7). After the war, Hogben entered academic life, teaching and leading research in London at Birkbeck and the Royal College of Science (1917–1922), in Edinburgh at the Animal Breeding Research Laboratory (1922–1925), in Montreal at McGill University (1925–1927), and at the University of Cape Town in South Africa (1927–1930).

In his early career at Birkbeck, Edinburgh, McGill, and Cape Town, Hogben was primarily devoted to experimental embryology and physiology. He studied amphibian metamorphosis with Julian Huxley and amphibian pigmentation with Frank Winton (Huxley and Hogben 1922; Hogben and Winton 1922a, 1922b, 1923). These investigations were interventionist in design, attempting to elucidate the mechanisms responsible for phenomena; for example, he isolated the role of the pituitary in the amphibian pigmentory effector system by surgically going through the roofs of frogs' mouths and removing various portions of the gland, then noting the associated lack of pigmentation. Hogben also helped create the Society for Experimental Biology and its accompanying *British Journal of Experimental Biology*, which still exists today as the *Journal of Experimental Biology* (Crews et al. 1923; Erlingsson 2009). Hogben, with these achievements, quickly rose up the ranks of the British biological elite, joining the likes of Haldane and Huxley. Influential British geneticist Cyril D. Darlington recalled after Hogben's death, "When I was very young, Galdane, Guxley, and Gogben (as the Russians called them), seemed to be the three Magi."[11]

It was Hogben's seven years (1930–1937) at the London School of Economics and Political Science (LSE) that produced his most lasting contributions to science and society. During these years he wrote *Mathematics for the Million* (1937) and *Science for the Citizen* (1938); both books were hugely successful, designed for lay readers to teach themselves practical mathematics and science just as Hogben taught himself at the public library years earlier. While at the LSE, he also attacked Britain's eugenics movement with a tenacity unmatched even by the standards of other anti-eugenicists of his day (Blacker 1952; Ludmerer 1972; Werskey 1978; Soloway 1990; Barkan 1991; Mazumdar 1992; Kevles 1995; Paul 1995, 1998). Sir (later Lord) William Beveridge, director of the LSE, sought to bridge the divide between the natural and the social sciences and so announced the search for a Chair of Social Biology in 1929 (Dahrendorf 1995; Renwick 2013). Fisher actually inquired about filling the position, even envisioning whom he would hire as subordinates.[12] But it was Hogben whom Beveridge ultimately invited to take the post. In his autobiography years later, Hogben recalled this vocational victory with glee, noting, "the brass hats of the Eugenics Society were already congratulating themselves on the prospect of one of their co-religionists getting the job" (Hogben 1998, 121). Hogben, however, only agreed to take the appointment after some reluctance, later explaining, "At that time human genetics was a morass of surmise and superstition. ... Should I prosper in the Herculean task of cleaning the Augean stables of human heredity, I should be contributing to the overdue disposal of a manure heap of insanitary superstitions" (ibid., 122). Ultimately it was one of Hogben's fellow "Magi" who convinced him to take on the responsibility. "Conversation with J. B. S. Haldane jerked me out of indecision concerning my fitness for the task" (ibid.). Hogben accepted the position and left Cape Town, joining the LSE in 1930.

"A Third Class of Variability"

Hogben was not the first to formulate anti-eugenic arguments. Psychologists argued that the intelligence tests hailed by eugenicists to distinguish the fit from the unfit were biased; anthropologists claimed that the categories upon which eugenicists put so much emphasis—class, race, and ethnicity—were cultural constructs, not biological ones; even Catholics criticized eugenics on religious grounds, claiming marriage was a moral matter, not a eugenic one (Kevles 1995). Hogben instead attacked the methodological foundations of eugenic science. His first full-fledged assault on eugenics came with the publication of his *Genetic Principles in Medicine and Social Science* (Hogben 1932).[13] "This book does not undertake to set down all that is known and has been surmised about human inheritance," Hogben admitted. Instead, it was the first step in his "Herculean task." He explained:

It is an attempt to separate the wheat from the tares, to indicate where a sound foundation of accredited data is available, to discuss what methods can be applied to the extremely elusive nature of the material with which the human geneticist deals, and to re-examine some of the biological concepts which have invaded other fields of inquiry in the light of modern advances in experimental genetics. (ibid., 9)

The underlying thread that guided the discussion was his persistent emphasis on the role the environment played in the development of organisms. When it came to "a sound foundation of accredited data," Hogben emphasized the importance of the nutritional environment contributing to diseases such as rickets (ibid., 64). And when cognitive developmental disorders were discussed, Hogben drew attention to the effect of birth order on the incidences of the traits (ibid., 99–103).

When it came to reexamining "biological concepts," Hogben claimed, "Genetical science has outgrown the false antithesis between heredity and environment productive of so much futile controversy in the past" (ibid., 201). Since every trait is the end product of an immensely complicated series of developmental reactions between the environment and the hereditary material, "Differences can be described as determined predominantly by hereditary or predominantly by environmental agencies if, and only if, the conditions of development are specified" (ibid., 98). Hogben, to drive this point home, pointed out that variation in a population arose from hereditary variation (emphasized by eugenicists), environmental variation (emphasized by anti-eugenicists), and an often-ignored *third class of variability*: that which "arises from the combination of a particular hereditary constitution with a particular kind of environment" (ibid.). This "third class of variability" became especially important when Hogben employed it to criticize Fisher the following year.

But in 1932, Hogben had not yet criticized Fisher, and in his review of *Genetic Principles,* Fisher welcomed Hogben's hiring at the LSE (Fisher 1932).[14] Fisher began, "[Hogben's] recent appointment as Professor of Social Biology at the London School of Economics gave the welcome assurance that his keenly analytic brain, and training in a severe experimental discipline, would be put to important service in the study of the biology of man" (ibid., 147). Compliments aside, Fisher then complained that Hogben's attention to "purely academic considerations" too often led to an exclusion of "aspects of more practical importance" (ibid.). In particular, Fisher criticized Hogben's tendency to demand experimental investigations into the genetic and environmental causes of development before intervening on those causes of variation (with, for example, eugenic policies). Fisher sarcastically noted:

Throughout the book, those who consider that the practical importance of the problem renders it urgent, will receive a disturbing impression that they are being asked to wait, in solemn hush, outside the laboratory door, until the Professor sees fit to announce that the ultimate truth has at last been revealed. (ibid., 147–148)[15]

"An Inherent Relativity in the Concepts of Nature and Nurture"

That same year, the medical faculty at the University of Birmingham invited Hogben to deliver their William Withering Memorial Lectures, and Hogben chose medical genetics as the theme of his lectures. Hogben, in preparation for the lectures, contacted Fisher, who was about to leave Rothamsted for University College London, succeeding Pearson in the Galton Chair of Eugenics. "Dear Fisher," Hogben wrote, "I am at present engaged in preparing a course of lectures in which I shall be dealing with your own contributions to the genetic theory of correlation. There is one point in your 1918 paper which worries me very much." Hogben was concerned about Fisher's talk of "cause." He asked, "When you speak of the contribution of hereditable and nonhereditable causes of variance in a population, what exactly do you mean? I often use the same form of words myself and lately I have been searching for a more explicit formulation of the problem." To explain the source of his concern, Hogben provided a quantitative example:

Suppose you say that 90 per cent of the observed variance is due to heredity, do you mean that the variance would only be reduced ten percent, if the environment were uniform? Do you mean that the variance would be reduced by 90 per cent, if all genetic differences were eliminated? Perhaps you will think the question silly; but if you could suggest an alternative form of words, it might help.[16]

Fisher responded the following day.

Dear Hogben, Your question is a very sound one. The point is this:-If the differential effects of environment and heredity are not correlated, i.e. if each genotype has an equal chance of experiencing with their proper probabilities, each of the available kinds of environment, then the variance is additive, and the statements you have are equivalent.[17]

Fisher took Hogben's question to be one concerning the correlation of heredity and environment, and so he answered Hogben's question with a discussion of heredity's "chance of experiencing" a particular environment. This, however, was not Hogben's target, and so Hogben took several days to construct a lengthy rebuttal. "Dear Fisher," he wrote, "I don't think you quite got the difficulty which I am trying to raise. It concerns an inherent relativity in the concepts of nature and nurture."[18] To clarify, Hogben introduced research by American embryologist Joseph Krafka (1920), who raised different strains of fruit fly (one called "low bar" and another called "ultra bar") at different temperatures and then counted the number of eye facets that each population developed on their compound eyes:

From Krafka's data you will see the following values for facet number are given at 15° and 25° C.

	Low bar	Ultra bar
15° C	189	52
25° C	74	25

Consider the elementary population with the following structure. The genotypes are low bar and ultra bar in equal numbers, equally distributed between two environments, namely an incubator at 15° C and one at 25° C. There is zero correlation between the distribution of environmental and genetic variables. Yet I cannot agree that the two statements "y per cent of the variance is due to environment," and "the variance would be reduced by y per cent if all differences of environment were eliminated," are equivalent nor that there is equivalence between the two statements "x per cent of the variance is due to heredity" and "the variance would be reduced by x per cent if there were no genetic differences."

The reason, Hogben pointed out, was that there was a "lack of singularity in the problem," since differences in heredity and differences in environment could be eliminated in any number of ways. "Let us abolish all differences of environment," Hogben proposed:

We can do this in an infinite number of ways. One would be to culture all flies at 15° C. Result: mean 120.5 and variance 4692. Another is to culture them all at 25° C. Result mean 49.5 and variance 600. Which of these two variances has priority as an estimate of the "contribution" of environment to the observed variance in the fourfold population? Again we eliminate all genetic differences by killing off all ultra bar flies. Result: mean 131.5 and variance 3306. We could alternatively kill off all low bar flies. Result: mean 38.5 and variance 182. Which of these gives the contribution of heredity to the observed variance?

Hogben's point was this: an analysis of variance that partitioned causes of variation in a population did not necessarily translate into a single answer about how the population would respond to an intervention on those causes of variation; it depended on how the hereditary difference was acted upon, and how the environmental difference was acted upon. Hogben closed by making the source of his concern explicit: "What I am worried about is a more intimate sense in which differences of genetic constitution are related to the external situation in the process of development."

Hogben's letter on the 23rd of February, 1933, marked the dawn of interaction being utilized as a critical tool to attack the summing of hereditary and environmental causes of variation.[19] The fruit fly example from Krafka also became the empirical backbone of Hogben's last William Withering lecture, entitled "The Interdependence of Nature and Nurture" (Hogben 1933a, 1933b). It was, in short, an all-out attack on Fisher.

There, Hogben admitted that Fisher's statistical techniques could be "used to detect the existence of differences due to environment and differences due to heredity" (Hogben 1933a, 93). However, moving beyond the detection of such differences, "The difficulties of interpretation begin when we attempt to clarify what is meant by calculating 'the numerical influence ... of the total genetic and non-genetic causes of variability'" (ibid., 94–95). Hogben drew on his Cambridge philosophical hero to make this point. "In his illuminating essay on the *Notion of Cause* Bertrand Russell has pointed out that few words are used with greater ambiguity in scientific discussion" (ibid., 95).[20] What Hogben had in mind here was an extension of the critique he first made in his *Genetic Principles*. "The biometrical treatment of variability," Hogben argued,

> inherited from Galton a tradition of discourse in which the ambiguity of the concept of causation completely obscured the basic relativity of nature and nurture. Since then this relativity has become increasingly recognised through experiments involving the use of inbred stocks in physiological laboratories, especially in connexion with experimental work on diet. It is therefore necessary to examine with great care what we mean when we make measurements of a genetic difference and a difference due to environment. (ibid.)

Hogben, to explain "what we mean" when experimental biologists talk about "genetic difference" and "environmental difference," introduced to his reader the same case he introduced to Fisher in correspondence earlier that year—Krafka's data on different strains of fruit fly raised at different temperatures. This time, though, Hogben provided both the data and an image depicting the different responses of the low bar and the ultra bar strains to the different temperatures (see figure 2.2). Figure 2.2 is a norm of reaction graph; it plots the differential response of different genetic groups (their "norm of reaction") to different environments. On such a norm of reaction graph, if the norm of reaction lines are parallel, then there is no interaction of nature and nurture; if the lines are not parallel, however, then there is interaction in the population. In figure 2.2, notice that the norm of reaction line for low bar and the norm of reaction line for ultra bar are clearly not parallel, indicating that there was interaction in that

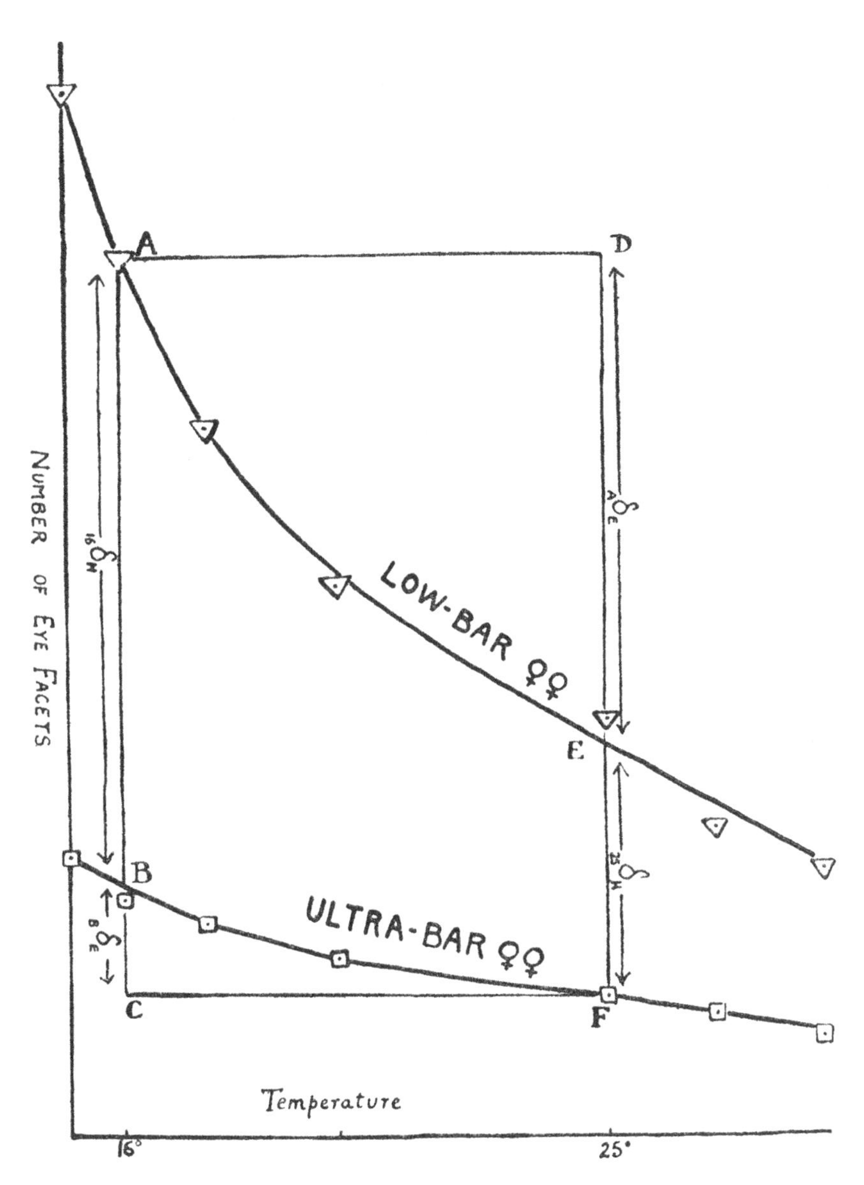

Figure 2.2
Hogben's norm of reaction graph for number of eye facets (y-axis) for low bar and ultra bar *Drosophila* strains raised at different temperatures (x-axis; reproduced from Hogben 1933a, figure 2).

population. The differences between points A and B and between points E and F corresponded to what Hogben claimed experimental biologists meant by a genetic difference. The differences between points B and C and between points D and E corresponded to what Hogben claimed experimental biologists meant by a difference due to environment. Hogben granted, "Clearly we are on safe ground when we speak of a genetic difference between two groups measured in one and the same environment or in speaking of a difference due to environment when identical stocks are measured under different conditions of development." But then he continued, questioning, "Are we on equally safe ground when we speak of the contribution of heredity and environment to the measurements of genetically different individuals or groups measured in different kinds of environment?" (ibid., 97). Hogben asked his reader to consider a low bar stock kept at 16° C and an ultra bar stock kept at 25° C, creating the observed differences AC or DF. "How much of AC or DF is due to heredity and how much to environment? The question is easily seen to be devoid of a definite meaning" (ibid.).

Yet it was precisely such meaning, Hogben claimed, that eugenicists attached to Fisher's analysis of variance (or what Hogben called Fisher's "balance sheet of nature and nurture"). Recall that Fisher, in 1918, wrote with regard to stature, "it is very unlikely that so much as 5 per cent. of the total variance is due to causes not heritable," to which Hogben retorted, "The only practical significance which Fisher's analysis of variance seems to admit is that, if it were correct, we could only reduce variance with respect to stature in a human population by 5 per cent. or less if the environment were perfectly uniform" (ibid., 114). But the lesson from the different strains of fruit flies raised at different temperatures, Hogben argued, applied equally well to different classes of humans raised in different medical, nutritional, and educational environments. The human environment could be made uniform in any number of different ways, some ways tending to magnify differences between classes and some tending to diminish differences between classes (ibid., 116). Hogben concluded,

> In whatever sense Fisher himself intended his balance sheet to be interpreted, there is no doubt that many writers on human biology entertain the belief that biometrical estimates of this kind do entitle us to set such limits. On the basis of such statements as the previous quotation about stature, it is often argued that the results of legislation directed to a more equitable distribution of medical care must be small, and that in consequence we must look to selection for any noteworthy improvement in a population. This is rather like saying that the difference between black and

white is negligible because an inkpot thrown into a tank of china clay has very little effect on the latter. (ibid., 116–117)

We can only assume that Fisher felt little gratitude when Hogben concluded his essay, "It is a great pleasure to acknowledge the courtesy with which Dr. Fisher has replied to communications in which some of the issues raised in this discussion have been explored" (Hogben 1933b, 405).

The Explanatory Divide

Although Hogben never mentioned it, Fisher responded to Hogben's letter discussing the Krafka data just days later:

Dear Hogben, I think I see your point now. You are on the question of non-linear interaction of environment and heredity. The analysis of variance and covariance is only a quadratic analysis and as such only considers additive effects. Academically one could proceed in theory, though in a theory not yet developed, to corresponding analyses of the third and higher degrees. Practically it would be very difficult to find a case for which this would be of the least use, as exceptional types of interaction are best treated on their merits, and many become additive or so nearly so as to cause no trouble when you choose a more appropriate metric. Thus facet number shows its sweet reasonableness when measured in 'proportional units' or in other words on a logarithmic scale. However perhaps the main point is that you are under no obligation to analyse variance into parts if it does not come apart easily, and its unwillingness to do so naturally indicates that one's line of approach is not very fruitful.[21]

Fisher's appraisal of interaction here reveals much about his understanding of the phenomenon by the time Hogben used it against him. Fisher saw Hogben now to be worrying about the " non-linear interaction of environment and heredity." Fisher, of course, was familiar with the problem, having taken up his study of potato varieties and different fertilizers in 1923 with the sole purpose of testing for the complication. With the conclusions of that study in mind, notice how Fisher responded to Hogben: his concern was written off as "academic," while "practically it would be very difficult to find a case for which this would be of the least use." Fisher's investigation at Rothamsted led him to believe that such cases of interacting causes of variation were not the norm in nature; and, if they did arise, they could be eliminated by altering the scale on which the environment was measured (a transformation of scale), such as with the fruit fly data's "sweet reasonableness" with a transformation to a logarithmic scale. Notice also that Fisher's letter bore a striking resemblance to his review of *Genetic Principles* a year earlier, in which he complained that Hogben's attention to "purely academic considerations" led to an exclusion of "aspects of more practical importance." Hogben could pursue a study of interactions for academic reasons, Fisher granted, but such a study would have no practical application.

Fisher's congeniality in this correspondence, which predated Hogben's published assault on Fisher, can be contrasted with a letter he wrote to fellow-eugenicist John A. Fraser Roberts in 1935 after Hogben published the last William Withering lecture, "The Interdependence of Nature and Nurture." There, Fisher predicted, "There is one point in which Hogben and his associates are riding for a fall, and that is in making a great song about the possible, but unproved, importance of non-linear interactions between hereditary and environmental factors. J.B.S. Haldane seems tempted to join in this."[22] Fisher here assessed the interaction of nature and nurture quite explicitly: it was of "possible, but unproved, importance." "Possible" because, as Fisher recognized in his potato/fertilizer study, the non-linear interactions would greatly complicate his summing of the causes of variation. But also "unproved" because Fisher found no such deviations from the summation formula.

Hogben came to quite a different conclusion. Krafka's research was a clear example, and Hogben took full advantage of its implications in his anti-eugenics lectures and publications. Importantly, Hogben emphasized that the Krafka data was not an isolated instance, writing, "The literature of experimental physiology is not wanting in examples of such divergent curves representing the measurement of a character and the strength of the environment" (Hogben 1933b, 385). He drew on the research of Norman Taylor (1931) and Frank Winton (1927), his former colleague and co-author from Edinburgh, who examined variation in the sinus beat of frogs exposed to different temperatures, and variation in the mortality rate of rats exposed to different levels of rat poison, respectively (Hogben 1933b, 385).[23]

The Variation-Partitioning and the Mechanism-Elucidation Approaches to Nature and Nurture

When it came to judging the empirical evidence of interaction in nature, Fisher and Hogben evaluated the situation quite differently. Where Fisher saw a phenomenon of "possible, but unproved importance," Hogben saw a phenomenon he took to be common in nature. Why? Why, that is, did Fisher and Hogben see the situation so differently?

Historians who have reflected on the divergent scientific views of Fisher and Hogben have tended to emphasize the religious and political positions that motivated Fisher's eugenics and Hogben's anti-eugenics. Fisher was a devout Anglican who published on the relationship between science and Christianity (Fisher 1955). Historians have highlighted how Fisher's religion supported his eugenics; the two domains were united in their emphases on family and in their emphases on human progress (Ruse 1996; Bartley 1994; Moore 2007). Fisher was also a proud patriot (eager to fight for his country) and a political conservative. He saw a clear divide between the upper and lower classes in Britain, and he thought eugenics could improve his country by increasing the former and decreasing the latter (Mazumdar 1992; Kevles 1995).

Hogben, in stark contrast, was an evangelical atheist. "I'm an atheist, thank God!" he often said (Wells 1978, 190). His perspective on religion aligned well with his socialist politics. He was critical of Britain's class system, a system which he spent his life demonstrating was no guarantee of performance. In 1931, at the Second International Congress on the History of Science, a Soviet delegation led by Nikolai Bukharin introduced Marxism to the British scientific community. Historians have noted how influential this visit was for left-wing scientists of the day, such as Hogben and Haldane. As Mazumdar explained:

> Hogben's thinking on the problems of social biology did not take a completely new direction following his contact with Marxism, but the Marxist analysis both sharpened his perception of the class-bound nature of the eugenic program, and also provided a theoretical support for his campaign against the over-emphasis of the biological in human society. (1992, 161)

Gary Werskey similarly offered:

> Rather than completely sacrifice [Hogben's] outside political interests to the demands of scientific life, he consciously brought his politics to bear on the kind of science he did. As a feminist who was also an experimental biologist, Hogben was drawn in the early twenties to the new field of comparative endocrinology, in order to study the hormonal bases of sex differences. As a socialist, he likewise found himself attracted to the social biology of class and racial differences. (1978, 105; see also Kevles 1995; Bowler 2001)

These appeals to religion and politics certainly help us to appreciate the motivations for Fisher's eugenics and Hogben's anti-eugenics. But the separate routes that Fisher and Hogben took to first considering the interaction of nature and nurture revealed something else that divided them. In addition to any religious or political divide, they also faced an explanatory divide. They faced this explanatory divide because, although they shared a common interest in understanding the relationship between nature and nurture, they employed very different approaches to studying that relationship. I use "approach" here in a manner similar to Thomas Kuhn's "paradigm" (1962), Imre Lakatos' "research programme" (1977), Larry Laudan's "research tradition" (1977), or Ian Hacking's "style of scientific theorizing" (1994). These various philosophical accounts emphasize how different scientists can attempt to explain an apparently common phenomenon and yet approach that phenomenon with very different questions, concepts, problems, and methodologies. My emphasis is on how Fisher and Hogben, though they studied a common phenomenon—the relationship between nature and nurture—approached that phenomenon with different explanatory frameworks. That is, though they were both interested in the relationship between nature and nurture, they identified different things that needed explaining (what philosophers call the "explanandum"), they asked different causal questions about the explanandum, they appealed to different things that did the explaining (what philosophers call the "explanans"), and they pointed to different methodologies that generated the explanans. Fisher and Hogben faced an explanatory divide because they diverged on each of these components: thing to be explained, causal question, thing that does the explaining, and methodology. Let me explain.

Fisher took what I call the variation-partitioning approach to studying nature and nurture (see table 2.1). For him, the thing that needed explaining was variation in a population or, more specifically, the relative contributions of nature and nurture to variation in a population. Fisher asked, how much of the variation was due to hereditary causes of variation, and how much was due to environmental causes of variation? What did the explaining when it came to these how-much questions were the causes of variation responsible for variation in a population. Fisher, in turn, developed many of the statistical methodologies, such as the analysis of variance, that allowed for partitioning these causes of variation in order to attribute to each its relative contribution.

Hogben, in contrast, took what I call the mechanism-elucidation approach to studying nature and nurture (see table 2.1). For Hogben, the thing that needed explaining was the developmental process or, more specifically, the developmental relationship between nature and nurture that played out during the developmental process. Hogben asked, how do differences in heredity and how do differences in environment interact during development to create variation separate from hereditary and environmental variation alone? What did the explaining when it came to these how questions were the causal mechanisms responsible for the

Table 2.1
The Components of the Explanatory Divide

	Variation-partitioning approach	Mechanism-elucidation approach
Thing to be explained	Variation in a population	Developmental process
Causal question	How much?	How?
Thing that does the explaining	Cause of variation	Causal mechanism
Methodology	Statistical	Interventionist
Concept of interaction	Biometric	Developmental

development of a trait. Hogben, for his part, thought Fisher's statistical methodologies were useful as a starting point for detecting the relevant variables in the causal mechanisms, but those statistical methodologies had to be followed up with more interventionist experiments designed to probe the system in order to elucidate how those variables made their difference in the developmental mechanisms.[24]

Different Answers to the Conceptual, Investigative, and Evidential Questions

Fisher and Hogben shared a common interest in studying the relationship between nature and nurture, but they faced an explanatory divide because they took very different approaches to that study. Fisher's variation-partitioning approach attended to the causes of variation responsible for variation in a population, while Hogben's mechanism-elucidation approach attended to the causal mechanisms responsible for the development of a trait. Their routes to first considering the interaction of nature and nurture were situated in these approaches. And because they approached the study of nature and nurture so differently, it followed that their considerations of interaction were different, too. Fisher and Hogben ultimately had very different answers to the following three questions:

- the conceptual question—what is interaction?
- the investigative question—why and how should interaction be investigated?
- the evidential question—what is the empirical evidence for interaction?

When it came to the conceptual question, Fisher defined interaction as a "deviation from summation" and also as a "non-linear interaction." For

Fisher, notice, the interaction of nature and nurture was defined as an *absence*—an absence of summation, an absence of linearity, an absence of additivity between the main effects. It was also a purely statistical concept, a product of his statistical analysis of variance. Interaction arose in an analysis of variance when the main effects of nature and nurture failed to add up to the total variation. Fisher conceptualized interaction as *a statistical measure of the breakdown in additivity between the main effects of nature and nurture*. Because, for Fisher, interaction was a statistical phenomenon, I call this the "biometric concept of interaction."

Fisher's biometric concept of interaction influenced his answer to the investigative question. Interaction had to be considered because it posed a "great complication" to the task of measuring the relative contributions of nature and nurture to variation in a population. Partitioning those relative contributions was quite easy if the causes of variation were independent. But if they were not independent, if the variation caused by heredity was affected by the variation caused by environment and vice versa, then the statistician and the eugenicist suddenly faced a "considerable hazard." Fortunately, as far as Fisher was concerned, there was a solution to the problem if it arose. If the analysis of variance did indeed find a non-additive deviation from the summation formula, then the statistician could alter the scale on which the environmental variable was measured (to, for example, a logarithmic scale) in order to eliminate the non-linear interaction and get back to assessing the relative contributions of the main effects.

Even more fortunate, as far as Fisher was concerned, the transformation of scale wasn't commonly needed for the simple reason that non-linear interactions were the exception and not the rule in nature. They were a "possible, but unproved" obstacle that he had considered and then dismissed in 1923 when he studied the different potato varieties' responses to different fertilizers. This was Fisher's answer to the evidential question. And it explains why he wrote off Hogben's concern as "academic." In nature, where scientists and eugenicists with "practical" concerns focused their attention, Fisher saw no reason to be concerned by the prospect of that great complication.[25]

Hogben, coming to consider interaction within a mechanism-elucidation approach, answered the conceptual, investigative, and evidential questions much differently. Hogben clearly did not get the original idea from Fisher. Hogben never referenced Fisher's 1923 publication on the deviations from summation; and, when Hogben first raised the issue with Fisher in correspondence, remember that he admitted, "Perhaps you will think the question silly." Hogben, in his letter to Fisher, explained quite clearly the origin of the issue for him: "What I am worried about is a more intimate sense in which differences of genetic constitution are related to the external situation in the process of development." This was his third class of variability—variation due to the combination of a particular hereditary constitution with a particular kind of environment. Hogben's answer to the conceptual question was thus much different from Fisher's answer. Interaction, according to Hogben, was not an absence; it was a *presence*—a presence of the third class of variability, a presence of a source of variation generated by the developmental relationship between nature and nurture. Hogben conceptualized interaction as *variation that resulted from differences in unique, developmental combinations of nature and nurture.* Because, for Hogben, interaction was a developmental phenomenon, I call this the "developmental concept of interaction."

Hogben's developmental concept of interaction influenced his answer to the investigative question. The interaction of nature and nurture was to be sought out, according to Hogben, because of the information it provided about the causal mechanisms of the developmental process. Statistical methodologies, such as Fisher's analysis of variance, were a useful starting point to "detect the existence of" hereditary differences and environmental differences, but then the scientist had to transition to interventionist experiments that elucidated how those differences made their difference in the causal mechanisms of development.[26]

Because, according to Hogben, the interaction of nature and nurture was a developmental phenomenon, it was to be commonly expected in nature. And this is why he answered the evidential question so differently from Fisher. There was Krafka's (1920) example, and Hogben added Taylor (1931) and Winton (1927), surmising, "The literature of experimental physiology is not wanting in examples." The interaction of nature and nurture, understood as Hogben conceptualized it, was the rule and not the exception.

In Fisher's last letter to Hogben, he began, "I think I see your point now. You are on the question of non-linear interaction of environment and heredity." The thesis of this chapter about the origin*s* of interaction, in the plural, is that Fisher was partially right but also partially wrong in this statement. There was a sense in which Fisher did "see" Hogben's point—he realized Hogben was concerned about the interaction of nature and nurture. There was also a sense, though, in which Fisher did not "see" Hogben's point—he did not conceptualize, investigate, or judge the empirical evidence of that interaction at all like Hogben. Part of the divide between

Fisher and Hogben was surely political. But I have argued here that there was also an explanatory divide between them. They answered the conceptual, the investigative, and the evidential questions so differently because they came to consider the interaction of nature and nurture from two very different approaches, each with its own thing to be explained, causal question, thing that does the explaining, and methodology.[27]

The Legacy of Fisher versus Hogben

The exchange between Fisher and Hogben evidently took its toll on their relationship. In 1932, when reviewing Hogben's *Genetic Principles*, Fisher welcomed Hogben's appointment to the Chair of Social Biology at the LSE. But in an unpublished review of Hogben's *Nature and Nurture*, where Hogben attacked Fisher, Fisher chided,

> Many of those, who had hopes that the establishment of a Chair of Social Biology at the London School of Economics would lead to a scientific and unbaissed [*sic*] attack on the social problems in this field, must by now be realising, in various degrees, their disappointment. For the functions of an advocate and of an investigator seem to be incompatible; and though one may be always amused and sometimes stimulated to thought when a brilliant journalist, such as Mr. G. K. Chesterton, sets out to show what a good forensic case can be made in opposition to the weight of scientific evidence and opinion, Professor Hogben lacks the charm of style needed to make confusion of thought seem luminous, or his facetiousness seem penetrating.[28]

Fisher's disdain for Hogben was by no means confined to the years of their debate. Almost thirty years later, when there was some confusion over whether an article in *Nature* was written by Anthony W. F. Edwards (Fisher's student) or his brother John Edwards (Hogben's student), Fisher wrote of the matter to his former University College London colleague R. R. Race:

> It was the thought that it was he [i.e., Anthony W. F. Edwards] that annoyed me, for the estimates published in *Nature* were manifestly incompetent, and I feared that one of my own pupils was running amok, and adding unnecessarily to darkness and confusion. However, I understand he [i.e., John Edwards] is only one of Hogben's, so all is explained.[29]

Fisher's Edwards, in fact, personally witnessed his mentor's disdain for Hogben upon the arrival of the paperback edition of Fisher's *The Genetical Theory of Natural Selection* (1958). "I was standing in the departmental office when Fisher opened the parcel of author's copies," Edwards recalled. "'Hmph,' [Fisher] said at his first sight of the cover, 'Looks like a book by Hogben'"[30] (Edwards 1990, 278).

Hogben lost no less love. In discussing the downfall of the Nazi Party in an unpublished portion of his autobiography, Hogben judged:

> After the war, the Nuremberg justices of the peace had [Nazi Alfred] Rosenberg hanged. If I believed in hanging people for their opinions, the only extenuating circumstances I might enter with a clear conscience as a plan for mercy on behalf of the late Sir R. A. Fisher would be that he did not occupy a government post with responsibility for implementing his convictions.[31]

That from an avowed pacifist!

Importantly, the debate between Fisher and Hogben did not just affect their personal relationship. Their separate routes to the interaction of nature and nurture generated two different concepts of interaction, two different assessments of why and how interaction should be investigated, and two different judgments of the empirical evidence for it. These different answers to the conceptual, investigative, and evidential questions outlived the debate between Fisher and Hogben and even outlived them. When the next debate over interaction erupted in 1969, Fisher was dead and Hogben was beginning his retirement in Glyn Ceiriog, Wales. And yet the explanatory divide arose again, this time surrounding the controversial question, "Why do African Americans perform poorer on IQ tests than Caucasians?"

3 Race, Genetics, and IQ: Interaction in the IQ Controversy

The debate between Fisher and Hogben discussed in the last chapter took place at the height of the American and British eugenics movement. It was an era of coerced sterilizations (Largent 2007), immigration restriction acts (Ordover 2003), and racist-inspired science (Jackson 2005), all justified in the name of eugenics (Kevles 1995). By the time Fisher and Hogben were reminiscing about each other with disdain in the 1960s, however, the scientific and social landscape had changed dramatically. After World War II, the world learned more about the atrocities enacted by the Nazis, who employed eugenic arguments to justify their actions (Kühl 1994; Weiss 2010). Genetic counseling was recast as an exercise in family decision making, rather than an exercise in the state's control of breeding (Lombardo 1996; Slavishak 2009). The simplistic one-gene-one-trait picture of genetics gave way to a much more complicated understanding of the relationship between nature and nurture (Paul 1998; Allen 2011). Eventually, the very term "eugenics" became stigmatized.

In the United States, the 1960s was also the decade of the African American civil rights movement. Instead of drawing clear lines between different races and acting on those differences with segregation, discrimination, and disenfranchisement, activists in the civil rights movement sought to break down that racial divide. Social organization became a powerful force, with boycotts, sit-ins, freedom rides, and marches used as non-violent tools of change (Dierenfield 2008). These efforts led to significant victories—public schools and universities were integrated (at times under the threat of federal military intervention), and federal legislation was passed to prevent discrimination and disenfranchisement, such as the Civil Rights Acts of 1964 and 1968 and the Voting Rights Act of 1965 (Crain 1968; Laney 2003; Loevy 1997). But the efforts also led to significant tensions—some Americans vehemently opposed these changes to the status quo; members of the Black Power movement argued that non-violent protests would never lead to true social change; urban unrest became common (Fine 2000; Gerstle 2002; Lumpkins 2008). In 1968, Martin Luther King Jr. was assassinated (Dyson 2008).

It was in this environment that an educational psychologist at Berkeley published "How Much Can We Boost IQ and Scholastic Achievement?" (Jensen 1969). Arthur Jensen noted that IQ had a very high heritability, suggesting that differences in IQ were largely the result of genetic differences. Since African Americans, on average, performed more poorly on IQ tests than Caucasians, Jensen argued that the racial gap for IQ could itself be explained genetically. Thus, egalitarian attempts to create an environment in which the racial gap for IQ disappeared were misguided; the gap was largely a genetic one and so would not disappear entirely with environmental intervention.

Unleashed at the height of the civil rights movement, Jensen's publication created a furor in academia, the media, and the general public. This was the birth of the "IQ controversy." Here was an academic from a prestigious American university who seemed to be offering a scientific foundation for treating the races differently, when the last decade had been spent combating such differential treatment. No academic was more forceful and influential in criticizing Jensen than Harvard evolutionary geneticist Richard Lewontin. Lewontin employed the interaction of nature and nurture (or "genotype-environment interaction" when he was writing) to refute the genetic hypothesis of race differences, just as Hogben had done thirty years earlier for British classes. Genotype-environment interaction was a common biological phenomenon, Lewontin warned, and hypotheses about differences between races could not be easily inferred when this interaction existed. Importantly, however, Jensen was unimpressed by the prospect of genotype-environment interaction. Just as Fisher dismissed Hogben, Jensen countered Lewontin: statistical analyses revealed no such interaction (properly understood), so there was no use dwelling on it.

How do we explain this scientific debate? Why did Jensen and Lewontin conceive of genotype-environment interaction and its implications for assessing racial differences so differently? An answer to this question has emerged from philosophers and historians of science: ignorance and sociopolitical bias. Two sides have arisen, which I will call the "Lewontinians" and the "Jensenites." Lewontinians have argued that Jensen (and defenders of Jensen) were ignorant of the statistical methods used by geneticists; ignoring genotype-environment interaction long after Lewontin highlighted the reality and importance of it could only be motivated by sociopolitical bias. The Jensenites have argued that Lewontin (and defenders of

Lewontin) were simply confused about the nature of genotype-environment interaction; making much ado about interaction long after Jensen pointed out this confusion could only be motivated by...well, sociopolitical bias.

This is a strange place indeed to find ourselves. As will be conveyed in the following pages, Jensen and Lewontin posited a variety of interrelated arguments for either dismissing or emphasizing genotype-environment interaction. Moreover, they both appealed to the existing empirical data on interaction at that time. And finally, Jensen and Lewontin were no idiots; they were both academics at premier universities with extensive statistical training and research. Thus, simply appealing to ignorance and sociopolitical bias to write off either position is much too quick. There must have been something more that divided Jensen and Lewontin. This chapter will explicate that "something more." In particular, I will argue that Jensen and Lewontin found themselves in the same predicament that Fisher and Hogben encountered in the last chapter; they faced an explanatory divide concerning how the interaction of nature and nurture should be conceptualized, investigated, and evaluated. This explanatory divide in turn led them to very different answers to the same questions that divided Fisher and Hogben:

- the conceptual question—what is interaction?
- the investigative question—why and how should interaction be investigated?
- the evidential question—what is the empirical evidence for interaction?

I want to be very clear from the beginning about what I am claiming and what I and am not claiming in this chapter. I am not claiming that my explication of the explanatory divide leaves no room for sociopolitical interests to play a role in the IQ controversy; the debate over race, race differences, and the cause of those differences was highly charged, and the disputants certainly had their eyes on the ethical, social, and political implications of the conclusions. What I am claiming, though, is that any account of this history that settles for characterizing the debate solely in terms of ignorance and sociopolitical bias is incomplete; the explanatory divide is a crucial part of this story, and so understanding this explanatory divide generates a more complete picture of why Jensen and Lewontin disagreed during the IQ controversy, and why the Jensenites and the Lewontinians have continued to disagree long after the IQ controversy. More importantly, the identification of an explanatory divide calls out for a philosophical resolution of that divide such as the one that I take up in the chapters of part II. A charge of ignorance or sociopolitical bias, in contrast, does the exact opposite; it shuts down further discourse. So in addition to a more complete historical picture, we are also rewarded with a redirected focus of this debate—a focus on a domain where philosophical analysis can continue to function.

The IQ Controversy

By the 1960s it was widely known among educational psychologists that there existed on average a gap in IQ scores between African Americans and Caucasians. This "IQ gap" amounted to about 15 points, more than one standard deviation. The IQ gap raised two, related questions. First, what explained the gap? And second, how could the gap be made to go away in a society that prided itself on egalitarian principles of equal opportunity? The standard answer to the first question invoked the very issues the civil rights movement was designed to abolish: segregation, discrimination, and disenfranchisement, which led to a systematically poorer educational environment for impoverished minority groups. With that explanation in place, the answer to the second question about eliminating the gap was a compensatory educational system designed to target impoverished minority groups with supplemental education to make up for the systematic disadvantages of their environment (Vinovskis 2005). The flagship compensatory educational program was (and remains) Head Start (Zigler and Styfco 2010). The program was created in 1965 as part of President Lyndon Johnson's more general war on poverty (Gillette 2010). Head Start was designed as a pre-kindergarten educational crash course of sorts wherein children from low-income families were exposed to an intense educational environment for several weeks before kindergarten in order to supply them with the social and cognitive resources to thrive in the kindergarten environment. After several years of compensatory educational programs such as Head Start, however, the IQ gap did not disappear. Why?

"Genetic Factors May Play a Part in This Picture"

By 1969, the field of behavioral genetics was taking shape as a unique discipline. The foundational text, *Behavior Genetics*, was published by John Fuller and William Thompson in 1960 (Fuller and Thompson 1960), and the Behavior Genetics Association along with its *Behavior Genetics* journal was established in 1970 (Fuller and Simmel 1986; Whitney 1990; Tabery and Griffiths 2010; Panofsky 2011; Schaffner forthcoming).[1] Behavioral geneticists applied the statistical methodologies of population genetics to study the relative contributions of nature and nurture to behavior in humans and non-human animals. In animals, behavioral geneticists often raised inbred strains of organisms such as mice in different environments in order to investigate how differences in genotype and differences in environment led to differences in traits (Lindzey and Thiessen 1970; Rader 2004; Nelson 2013; Leonelli et al. forthcoming). In humans, though, behavioral geneticists could not perform such experiments for obvious ethical reasons. Instead, they followed Galton's suggestion and examined the relative contributions of nature and nurture to behavioral traits by studying twins, adoptees, and other family relations (Vandenberg 1966). By studying identical twins reared apart or comparing adoptees to their biological and adoptive parents, for example, these human behavioral geneticists utilized statistical methodologies to partition the total phenotypic variation of some behavioral trait (V_P) into the genetic cause of variation (V_G) and the environmental cause of variation (V_E). Recall from chapter 2 that if these two main effects fully accounted for the total variation in the population (if they were "additive"), then the analysis of variance lined up quite neatly:

$$V_P = V_G + V_E.$$

Behavioral geneticists were particularly interested in how much of the total phenotypic variation was due to genetic differences. This genetic proportion was called the *heritability* of the trait:[2]

$$\text{heritability} = V_G/V_P.$$

Heritability was thus commonly understood as the "coefficient of genetic determination" (Meredith 1968, 337), the "degree of genetic determination" (Falconer 1969, 126), or the "index of the relative importance of gene differences" (McClearn and DeFries 1973, 201). Throughout the 1960s, behavioral geneticists reported the heritability of a variety human traits, ranging from schizophrenia to personality (Gottesman and Shields 1967; Nichols 1969).[3]

Arthur Jensen applied this human behavioral genetics approach to IQ. Jensen (1923–2012) spent most of his life at the University of California, Berkeley. He earned his BA in psychology there in 1945 and returned in 1958 as a professor of educational psychology, where he worked in emeritus status until his death in 2012. From 1956 to 1958 Jensen was a postdoctoral fellow with German psychologist Hans Eysenck at the Institute of Psychiatry, where he learned the behavioral genetic approach to studying personality and intelligence (Miele 2002).

Jensen titled his 1969 paper "How Much Can We Boost IQ and Scholastic Achievement?" He was asking: how much of the variation in IQ was due to the genetic cause of variation, and how much to the environmental cause of variation? Intervening on the environment with compensatory education would eliminate the IQ gap, his argument went, only if a large portion of the variation in IQ was due to environmental variation. If differences in IQ were instead mostly the result of genetic variation, then such environmental intervention was futile. So it was the heritability of IQ, Jensen argued, that would decide the situation. To calculate the heritability of intelligence, Jensen relied largely on the twin data of Sir Cyril Burt, Eysenck's mentor (Burt 1955, 1958, 1966). Based on that data, Jensen put the estimate for the heritability of intelligence at roughly 0.8, meaning that 80% of the variation in intelligence was accounted for by the genetic cause of variation or, in other words, 80% of the differences in IQ resulted from genetic differences (Jensen 1969, 51).

Unfortunately, Jensen bemoaned, "the possible importance of genetic factors in racial behavioral differences has been greatly ignored, almost to the point of being a tabooed subject, just as were the topics of venereal disease and birth control a generation or so ago" (ibid., 80). Jensen's goal was to break down this taboo. "There is an increasing realization among students of the psychology of the disadvantaged," Jensen surmised, "that the discrepancy in their average performance cannot be completely or directly attributed to discrimination or inequalities in education. It seems not unreasonable, in view of the fact that intelligence variation has a large genetic component, to hypothesize that genetic factors may play a part in this picture" (ibid., 82).

Jensen's controversial genetic hypothesis for race differences found favor with a number of notable scientists, such as his mentor Eysenck, Nobel laureate William Shockley, and Harvard psychologist Richard Herrnstein (Eysenck 1971; Herrnstein 1971; Shockley 1972). Herrnstein, in particular, popularized Jensen's hypothesis for a wider audience. In a 1971 article for *Atlantic Monthly*, Herrnstein introduced his readers to the research of Francis Galton and then Jensen (Herrnstein 1971). Far from being "extreme in position or tone," Herrnstein claimed Jensen's article simply summarized what was already widely recognized in scientific communities. "Not only its facts but even most of its conclusions are familiar to experts," Herrnstein wrote; "Jensen echoes most experts on the subject of I.Q. by concluding that substantially more can be ascribed to inheritance than to environment" (ibid., 55).

The Argument from Interaction

Many commentators on Jensen's genetic hypothesis, however, found his 1969 paper to be very much extreme in position and tone. And in response, critics attacked Jensen from a variety of angles. Burt's data, upon which Jensen initially relied to calculate the heritability of IQ, became an object of suspicion and seems to have been largely fabricated (Kamin 1974).[4] Compensatory education, defenders replied, had not failed; rather, it had not been given the correct formulation to allow it to succeed at its goals (Hunt 1969). IQ tests did not measure intelligence; they only measured how well people performed on IQ tests (Liungman 1974).[5] Richard Lewontin, however, sought to undermine the very methodological foundation of Jensen's genetic hypothesis.[6]

Lewontin (1929–) did his graduate work under the Ukrainian geneticist Theodosius Dobzhansky at Columbia University in the early 1950s.[7] Throughout the 1970s, Lewontin mixed his science and politics. He drew on his extensive training in evolutionary genetics to attack scientists who offered up biological explanations for the order of things, and he rallied like-minded scientists with the left-wing organization Science for the People (Lewontin, Paul, Beatty, and Krimbas 2001). Edward O. Wilson, author of the popular *Sociobiology* (1975), was one such target. Wilson argued that many complex human behaviors and social structures could be explained as products of natural selection based upon the evolutionary advantage such behaviors and structures offered a population. In reply, Lewontin criticized what he took to be an oversimplified picture of evolution and genetics, a picture in which simple genetic determinism replaced the biological reality of complex organism-environment interactions (Lewontin 1976). Jensen was another target, and Lewontin similarly saw in Jensen a naïve genetic determinism where complicated interactions were the biological reality.

"A Misapprehension about the Fixity of Genetically Determined Traits"

Lewontin began his assault on Jensen in 1970, right on the heels of Jensen's controversial 1969 publication. While Jensen took the importance of genetic factors affecting intelligence to create a problem for attempts at environmental intervention on the trait (i.e., compensatory education), Lewontin argued that a genetic component to intelligence in no way created such a problem. "Let it be entirely genetic," Lewontin granted (Lewontin 1970a, 8). "Does this mean that compensatory education, having failed, must fail? The supposition that it must arises from a misapprehension about the fixity of genetically determined traits" (ibid.). Genetic diseases, Lewontin pointed out, were once thought incurable because they were genetic. In the normal range of environments bearers of a genetic mutation would in fact develop their genetic disease. However, once the causal mechanisms responsible for disease development were understood, scientists were able to identify new environments that prevented the disorder from developing by, for example, altering nutritional intake. Jensen claimed that, because of IQ's high heritability, an environment of abundance would do little to elevate the lower IQ scores of African Americans in relation to Caucasians; Lewontin countered, "It is empirically wrong to argue that if the richest environment experience we can conceive does not raise I.Q. substantially, that we have exhausted the environmental possibilities" (ibid.). Determining the environments available to individuals, Lewontin emphasized, was a social matter, not a biological one. Thus, "In answer to Prof. Jensen's rhetorical question 'How Much Can We Boost IQ and Scholastic Achievement?' I say 'As much or as little as our social values may eventually demand'" (Lewontin 1970b, 25).

In this first critique of Jensen's genetic hypothesis, Lewontin introduced two key points that arose time and again in his subsequent discussions: (a) the importance of seeking the causal mechanisms of development when attempting to explain how the genotype and the environment contribute to a trait or differences in that trait, and (b) the importance of emphasizing possible, as-yet-untested environments as a source of new outcomes. Lewontin turned to genotype-environment interaction to develop these points throughout the 1970s as he challenged Jensen and the defenders of the genetic hypothesis of race differences.

In his influential "The Analysis of Variance and the Analysis of Causes," Lewontin (1974) utilized norms of reaction, just as Hogben did when attacking Fisher.[8] To recall from the previous chapter, a norm of reaction graph plots some phenotypic trait (P, along the y-axis) as a function of the environment (E, along the x-axis) and two (or more) different genetic groups; if the norms of reaction are parallel, then that means there is no interaction in the population, but if the norms of reaction are not parallel, then that means there is interaction. Lewontin introduced figure 3.1 to depict the phenomenon of developmental canalization, made famous by Conrad Hal Waddington. Waddington (1957) found that in a certain range of environments (the zone of canalization), two genetic groups responded almost identically (the area where the lines overlap significantly in figure 3.1). However, outside that zone, the genetic groups diverged quite

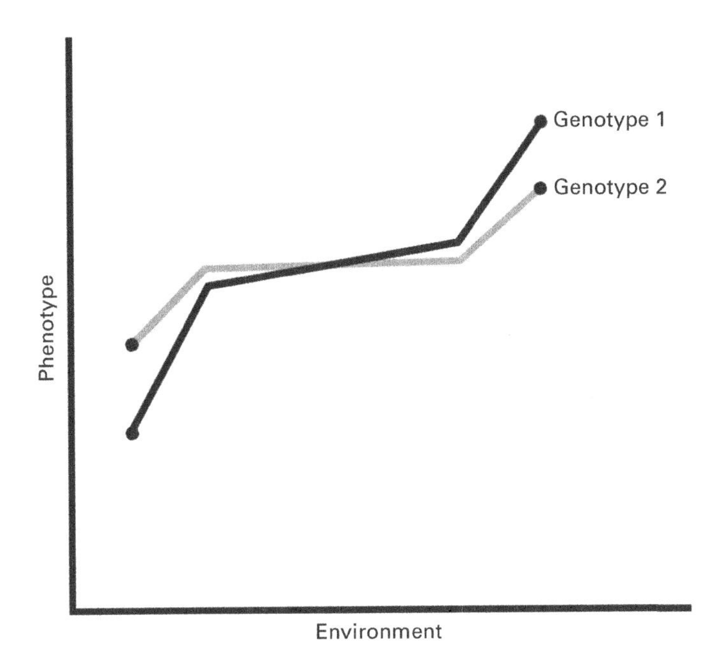

Figure 3.1
Lewontin's hypothetical norm of reaction graph for phenotypic traits caused by canalization (reproduced from Lewontin 1974, figure 1f).

dramatically. This zone of canalization in the normal range was a major insight of Waddington's developmental genetics, and Lewontin recognized it as a phenomenon that problematized Jensen's attempt to generalize from heritability measures. If interaction existed in a population, then the total variation (V_P) for the population could no longer be simply accounted for by the genetic cause of variation (V_G) and the environmental cause of variation (V_E); instead, an additional source of variation due to the genotype-environment interaction ($V_{G\times E}$) must be added as well:

$$V_P = V_G + V_E + V_{G\times E}.$$

Now, with this scenario, it became harder to derive a heritability measure for the trait under investigation because, rather than being additive and distinct, the genetic and environmental causes of variation were interdependent. Genotype-environment interaction thus posed a serious complication for anyone interested in providing a heritability measure for a trait.

The complex interdependence between the genetic and environmental distributions in figure 3.1 can be contrasted with those Lewontin provided in figure 3.2. Notice that the norms of reaction there are virtually parallel;

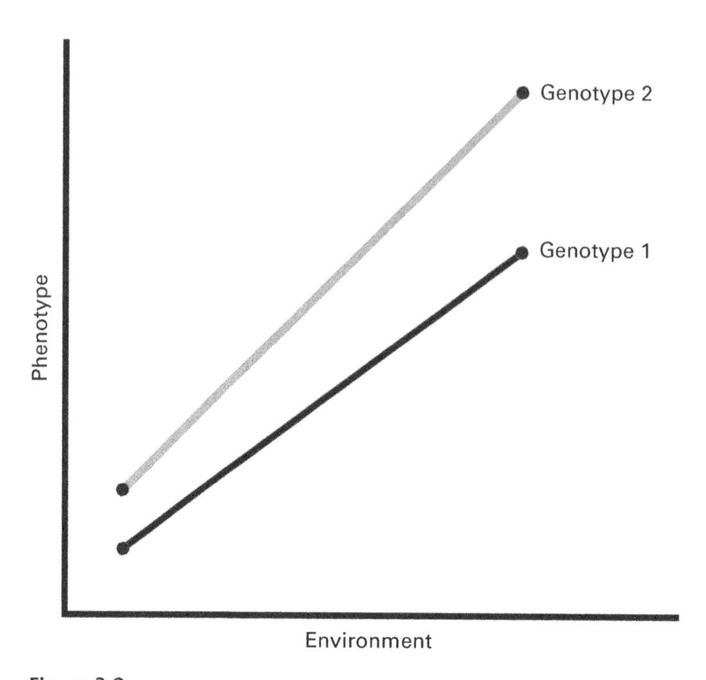

Figure 3.2
Lewontin's hypothetical norm of reaction graph depicting Jensen's genetic hypothesis for race differences in IQ (reproduced from Lewontin 1974, figure 1g).

there is very little to no genotype-environment interaction in figure 3.2. The question, then, was whether anyone could assume a trait like IQ looked like figure 3.2. Lewontin argued there was no scientific basis for such an assertion, and so additivity between the main effects was often simply assumed "because it suits a predetermined end" (Lewontin 1974, 409). This was the suspicious reasoning Lewontin ascribed to Jensen; he described figure 3.2 as "the hypothetical norm of reaction for IQ taken from Jensen (1969). It purports to show the relation between environmental 'richness' and IQ for different genotypes" (ibid.). Lewontin ridiculed such a picture. "While there is not a scintilla of evidence to support such a picture, it has the convenient properties that superior and inferior genotypes in one environment maintain that relation in all environments, and that as environment is 'enriched,' the genetic variance (and therefore the heritability) grows greater" (ibid.). Lewontin concluded sarcastically, "This is meant to take care of those foolish egalitarians who think that spending money and energy on schools generally will iron out the inequalities in society" (ibid.).

Lewontin reiterated these points a year later with Marcus Feldman in their criticism of behavioral geneticists' "Heritability Hang-up" (Feldman and Lewontin 1975). They warned, "If varieties are tested under a particular range of conditions, or a selection program is carried out over a limited range of environments, the selected material may be totally inappropriate for other conditions"—a direct reference to the problem posed by interaction (ibid., 1164). Importantly, Feldman and Lewontin charged Jensen and Herrnstein with ignorance of this interaction and its implications for their heritability estimates. They scolded, "This situation is ignored by both Jensen (1969) and Herrnstein (1971), whose discussion does not take account of this possible form of genotype-environment interaction" (ibid). Rather than focusing on heritability estimates, Feldman and Lewontin encouraged investigating the causal mechanisms responsible for the development of a trait:

> Relations between genotype, environment, and phenotype are at base mechanical questions of enzyme activity, protein synthesis, developmental movements, and paths of nerve conduction. We wish, both for the sake of understanding and prediction, to draw up the blueprints of this machinery and make tables of its operating characteristics with different inputs and in different milieus. (Feldman and Lewontin 1975, 1167–1168)[9]

Lewontin drew on the concept of genotype-environment interaction with norm of reaction graphs to make clear the importance he placed on the need to consider the causal mechanisms of development and possible, as-yet-untested environments in any discussion of variation and group differences. Interaction was then used, in turn, to attack the defenders of statistical analyses of variance, like Jensen, for their ignorance of such causal mechanisms and their confinement to limited, tested environments. Lewontin concluded his 1974 paper:

> The simple analysis of variance is useless for these purposes and indeed it has no use at all. In view of the terrible mischief that has been done by confusing the spatiotemporally local analysis of variance with the global analysis of causes, I suggest that we stop the endless search for better methods of estimating useless quantities. There are plenty of real problems. (410)

"The Phenotype Is a Developmental Process"

Importantly, Lewontin's project was not entirely negative, devoted only to tearing down the heritability measures utilized by Jensen and other behavioral geneticists. He also offered a positive vision for what geneticists ought to be studying. Lewontin emphasized the importance of research that examined the developmental relationship between nature and nurture, which could be visually represented with norm of reaction graphs. "The real object of study both for programmatic and theoretical purposes is the relation between genotype, environment, and phenotype," Lewontin wrote in 1974. "This is expressed in the *norm of reaction*, which is a table of correspondence between phenotype, on the one hand, and genotype-environment combinations on the other" (author's emphasis, 1974, 404). As an example of such a norm of reaction approach, Lewontin referred his readers to the early work of his mentor, Theodosius Dobzhansky. Dobzhansky's study of different flies' viability (y-axis) developing at different temperatures (x-axis) provided the norms of reaction graph found in figure 3.3, which also importantly offered for Lewontin an empirical instance of genotype-environment interaction (Dobzhansky and Spassky 1944). Lewontin even made use of the norm of reaction when contacting his mentor on Dobzhansky's 75th birthday, writing, "If there were a God, I would thank Him for bringing you so brilliantly through three-quarters of a century. As it is, you have only your genes and your environment to thank. With such a norm of reaction I have no doubt that you will reach 100!"[10]

Despite the fundamental importance Lewontin placed on the developmental relationship between nature and nurture, he was simultaneously concerned that perhaps he and Dobzhansky were among the only geneticists who properly recognized and emphasized this importance at the time of the IQ controversy. Writing to Dobzhansky in May 1973, Lewontin worried,

> you remain the *only* geneticist writing on general subjects and even one of the very few writing on technical subjects who says correct things about environment and genotype. The notion of the norm of reaction has simply failed to permeate the general textbook writings of our colleagues. As a result, they give all the wrong impression. ...Why is it that most geneticists do not understand that the phenotype is a developmental process?[11]

Lewontin's emphasis on interaction and its implications for heritability estimates was not confined to scholarly journals. In a 1973 article for the *Boston Phoenix*, an arts and entertainment weekly newspaper, journalist Paul Wagman covered the IQ controversy, reporting:

> Only people who have made an intensive study of quantitative genetics, says Lewontin, are prepared to understand the subject of heritability well enough to make such estimates. [David] Layzer, in a paper he has prepared for *Science*, maintains that the analyses which have led to the consensus cited by Herrnstein are shot through by systematic errors. (Wagman 1973, 28)

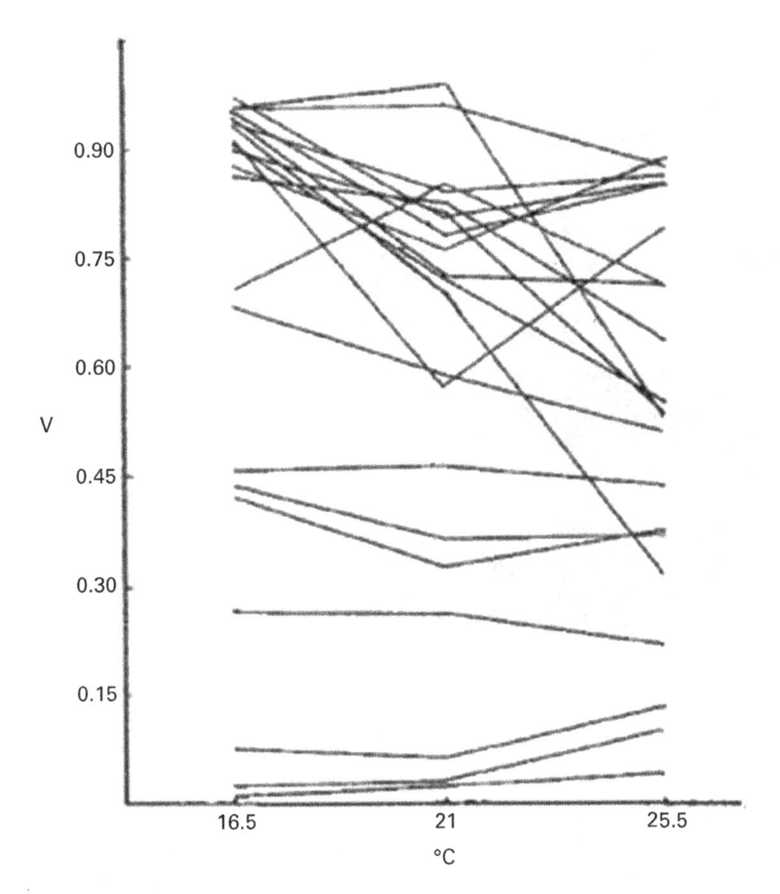

Figure 3.3
Lewontin's norm of reaction graph for the viability of different fruit fly strains raised at different temperatures (reproduced from Lewontin 1974, figure 2).

Figure 3.4
Photograph of Richard Lewontin from 1973 *Boston Phoenix* article "The Brains Do Battle in I.Q. Controversy" (with permission of Ken Kobre).

A photograph accompanied Wagman's report in which Lewontin seemed to jump off the page in frustration with his opponents (figure 3.4). And again genotype-environment interaction figured prominently in Lewontin's frustration; he stood adjacent to a blackboard on which norms of reaction were drawn for three genetic groups. Not surprisingly, the graph displayed significant interaction: Genotype 1 was superior to both Genotype 2 and Genotype 3 in environments to the left side of Lewontin's graph, but in environments to the right side of the graph, Genotypes 2 and 3 climbed high above Genotype 1. Genotype-environment interaction was thus the primary take-home lesson for even a journalist and photographer covering the IQ controversy.

"Such Characters Usually Reflect a Complicated Developmental Process"

Behind Lewontin's head in the photograph above is Harvard cosmologist David Layzer's name highlighted prominently, and in his reporting of the IQ controversy, Wagman united Lewontin and Layzer as the prominent critics of Jensen and Herrnstein. It was not surprising that Lewontin would mention his colleague at Harvard when emphasizing the importance of considering genotype-environment interaction in discussions of heritability estimates, for Layzer also drew on the concept of interaction to criticize Jensen in a number of articles throughout the early 1970s. Layzer first took up the matter in an exchange with Jensen on the pages of the first volume of the journal *Cognition* in 1972. Like Lewontin, Layzer drew attention to the assumed additivity between genotype and environment implemented by heritability estimates. He warned, "The assumption that genetic and environmental factors contribute additively and independently to a phenotypic character is, on general grounds, highly suspect" (Layzer 1972, 275). Layzer was especially critical of assuming additivity when it came to complex biological traits; "such characters usually reflect a complicated developmental process in which genetic and environmental factors are inextricably mingled" (ibid.).

Wagman's report on the IQ controversy mentioned a paper Layzer prepared for *Science*, which revealed that the consensus Herrnstein and Jensen pointed to in support of their genetic hypothesis was "shot through with systematic errors" (Wagman 1973, 28). Layzer's "Heritability Analyses of IQ Scores: Science or Numerology?" came out a year later, the same year as Lewontin's "The Analysis of Variance and the Analysis of Causes." Layzer also followed in the footsteps of Hogben by utilizing a norm of reaction graph to emphasize how analyses of variance, confined to limited environments, could miss important information about variation if genotype-environment interaction existed (figure 3.5). For instance, if a geneticist

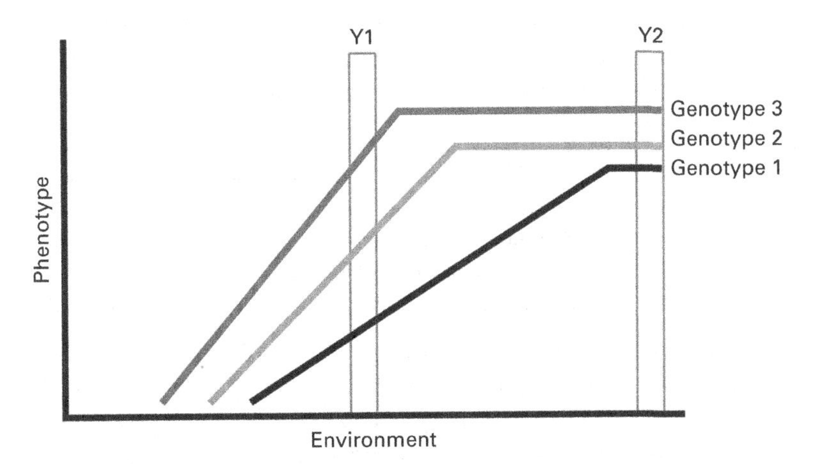

Figure 3.5
Layzer's hypothetical norm of reaction graph for three genotypes exposed to a variable environment (adapted from Layzer 1974, figure 1). (Reprinted with permission from AAAS)

only had information about the three genetic groups in the environmental region of Y1, then a heritability score would look much different than if the geneticist only had information about the genetic groups in the environmental region of Y2.[12]

Layzer's project, like Lewontin's, was also not entirely negative. Layzer too offered a positive thesis about what geneticists ought to be studying. And like Lewontin, Layzer highlighted the studies of Waddington, who focused on the developmental relationship between nature and nurture. Layzer used Waddington's research program as a metric by which to denigrate Jensen and Herrnstein's program. Waddington's views, he wrote, "contrast sharply with those of Jensen and Herrnstein, who believe in the possibility of discovering meaningful relations between measurable aspects of human behavior without inquiring into the biological or physiological significance of that behavior" (Layzer 1972, 273; see also Layzer 1974, 1260).

The Argument from Interaction Dismissed

Jensen, for his part, was quick to point out that he had considered the interaction of nature and nurture in his assessment, even if his critics accused him otherwise (Jensen 1976). And indeed, Jensen did take up the matter of genotype-environment interaction in 1969, although only to brush aside the complications the concept might have raised for his argument. There, Jensen lamented, "There is considerable confusion concerning the meaning of interaction in much of the literature on heredity and intelligence" (Jensen 1969, 39). He was critical of the growing group of "interactionists" whom he claimed were merely masked environmentalists, writing,

> Those who call themselves "interactionists," with the conviction that they have thereby either solved or risen above the whole issue of the relative contributions of heredity and environment to individual differences in intelligence, are apparently unaware that the preponderance of evidence indicates that the interaction variance ... is the smallest component of the total phenotypic variance. (ibid)

Jensen, at this early time, was already anticipating criticisms of his genetic hypothesis with appeals to interaction, which Jensen abbreviated as "V_I." "The magnitude of V_I for any given characteristic in any specified population is a matter for empirical study, not philosophic debate. If V_I turns out to constitute a relatively small proportion of the total variance, as the evidence shows is the case for human intelligence, this is not a fault of the analysis of variance model. It is simply a fact. If the interaction variance actually exists in any significant amount, the model will reveal it" (ibid., 41).

In his short discussion of genotype-environment interaction in 1969, Jensen introduced in an early form the three basic arguments he would continue to employ when discussing interaction throughout the IQ controversy: (a) that invocations of "interaction" were often simply confused about the meaning of the concept, (b) that discussions of interaction must be based on empirical data and not on abstract speculation, (c) and that if interaction existed for a human trait, then the analysis of variance would reveal the interaction, but the analyses of variance revealed no such interaction.

"A Failure to Understand the Real Meaning of the Term 'Interaction'"

Jensen employed the first of these arguments to attack any invocation of development in discussions of genotype-environment interaction. Under a section entitled "The Meaning and Non-meaning of 'Interaction,'" Jensen (1973) again grappled with his so-called "interactionists."

> Thus the interactionist theory holds that although there may be significant genetic differences at the time of conception, the organism's development involves such complex interactions with the environment that the genetic blueprint, so to speak, becomes completely hidden or obscured beneath an impenetrable overlay of environmental influences. (ibid., 49)

Jensen explained that this interactionist position "has arisen from a failure to understand the real meaning of the term 'interaction' as it is used in population genetics; but even more it is the result of failure to distinguish between (a) the *development* of the individual organism, on the one hand, and (b) *differences* among individuals in the population" (author's emphasis, ibid). Thus, any discussions of genotype-environment interaction that drew on the complexities of individual development were simply confused, and so there was no reason to engage with the arguments further (see also Plomin, DeFries, and Loehlin 1977 for a similar point from behavioral geneticists during the IQ controversy about confusions concerning interaction).

Jensen employed the second argument against abstract speculation when the inherent locality of the analysis of variance and heritability estimates were criticized. "The methods of biometrical genetics, of course, have no power to predict [heritability] under as yet untried interventions in the internal or external environments," he admitted when replying to "misconceptions about heritability" (Jensen 1975, 173). "It does give an indication of the relative influence of *existing* environmental sources of variance, and if [heritability] is very high, it tells us that merely reallocating individuals in existing environments will not have much effect in the rank ordering of individual differences" (author's emphasis, ibid). So the focus, Jensen claimed, was on the actual environments in existence and not on what the heritability of a trait might be in possible, as-yet-untested environments. Jensen drew on this point when replying to Lewontin's first attack. Lewontin claimed that compensatory education could not be written off as a failure simply because it proved unsuccessful in the existing educational system because new environments could be encountered or created that would facilitate such compensation. To this Jensen countered, "Lewontin seems to believe that anything is possible, given sufficient technological implementation. But reality does not bow to technology. Technology depends upon a correct assessment of reality" (Jensen 1970, 20; see also Plomin and DeFries 1976 for a similar point from behavioral geneticists during the IQ controversy about the need to focus on actual environments rather than possible ones).

When it came to the existence of genotype-environment interaction in actual environments, Jensen claimed that the analysis of variance was perfectly capable of detecting any such interactions, and so the fact that twin studies had not identified any such variation due to interaction in humans revealed the lack of any such variation in nature (Jinks and Fulker 1970). Responding to Layzer's commentary, Jensen noted, "Layzer makes much of the possibility of interaction of genetic and environmental factors" (Jensen 1972, 435). But then he continued, "In reading Layzer, one might easily get the impression that there is a lot of G×E interaction but that our models are unsuited to detecting it" (ibid.). To which Jensen retorted, "Not so."[13]

With these arguments in mind, Jensen ultimately placed the burden of proof on those who wanted to emphasize interaction's importance. "If G×E interaction is held up as a criticism or limitation of the applicability of heritability analysis to mental test data, the burden of demonstrating the presence of substantial G×E interaction in such data must be assigned to the critics" (Jensen 1975, 182).

Interestingly, Lewontin was not the only IQ controversy disputant corresponding with Dobzhansky during the episode. Jensen and Dobzhansky wrote to each other throughout the 1970s about the heated exchange between Jensen and Dobzhansky's former pupil. Dobzhansky, at one point, wrote Jensen in January 1972 to invite him to visit his laboratory, suggesting, "We could perhaps have an instructive discussion with you, of course in private. If we cannot convince each other, at least we could better understand each other."[14] And several years later, after Layzer's *Science* article was published, Jensen wrote Dobzhansky, complaining, "It is shocking to me that the article by Layzer should have been published in *Science* in its present form, because it is actually quite superficial and full of factual and technical errors which will greatly mislead readers who have not thought through these issues." Jensen even speculated as to the origin of Layzer's predilection in the debate: "I suspect that he is probably reluctant to pursue his mathematical inquiry (of which he certainly must be capable) to the point where it might begin to conflict with his obviously strong ideological preconceptions."[15]

Explaining the Controversy, Take 1: Ignorance and Bias

Lewontin and Jensen could not have been further apart in their assessments of genotype-environment interaction during the IQ controversy. On the one hand, Lewontin took such interaction to be of fundamental importance for understanding the nature of how a trait like IQ developed and, in turn, how variation in IQ emerged. The causal mechanisms of the developmental process had to be understood if scientists were going to propose changes to the educational system that could eliminate the IQ gap. On this understanding, possible environments were particularly insightful because they could reveal potential variation not already manifest in the existing environments. For Lewontin, the problem with users of heritability measures was that they ignored genotype-environment interaction simply because of the problems it created for the statistics, and ignored possible environments simply because the statistics needed to rely on data from actual environments. Jensen, on the other hand, took genotype-environment interaction to be little more than a dismissible nuisance. He did not ignore interaction, he replied. Interaction properly understood was not a developmental phenomenon, it was a statistical phenomenon—the absence of additivity. And Fisher's analysis of variance would detect that statistical phenomenon if it existed. But it did not. His critics' fixation on possible environments and the possible presence of genotype-environment interaction, Jensen retorted, was a distraction from the task at hand of evaluating the IQ gap in our actual environment.

How should we make sense of this divide between Lewontin and Jensen? What explains it? An answer to this question has emerged from a number of philosophers and historians of science who have investigated the IQ controversy; however, it is a peculiar one. Rather than drawing on the unique tools of the philosopher and the historian to shed new light on the episode, commentators have largely settled for simply reiterating the arguments of Lewontin and Jensen. Such appeals to authority then raise their own question: if either Lewontin or Jensen was obviously right (and the opposing disputant obviously wrong), then why did the disputant, along with defenders of the disputant, not see the obvious? Answer: ignorance and sociopolitical bias.

Lewontinians have largely populated the playing field since the time of the IQ controversy. Philosophers and historians began drawing on Lewontin's 1974 paper on the analysis of variance, in particular, immediately after it was published. Articles with hypothetical norm of reaction graphs showing significant genotype-environment interaction, emphases on possible environments, condemnations of heritability measures for their inability to elucidate the causal mechanisms of development, and recommendations to instead study norms of reaction quickly emerged, with due homage to Lewontin (Daniels 1974; Block and Dworkin 1976; Eckberg 1979). Michael Matthews, for example, suggested replacing heritability measures with norm of reaction research, arguing, "An important consequence of this change to norms-of-reaction research is that the analysis of variance, that time-honoured pursuit of IQ researchers, has next to no value" (Matthews 1980, 146). Matthews then turned to the appeal to authority to solidify his case. "Richard Lewontin traverses this terrain of population genetics and its putative connections with IQ research, and concludes: 'The simple analysis of variance is useless for these purposes and indeed has no use at all.'" (ibid., 146–147). In a similar vein, although in deference to Layzer, Allan Chase surmised,

> Professor Layzer, whose skills in mathematics and scientific logic and whose background and training in biology and genetics are certainly at least equal to those of Jensen and Herrnstein, concluded after a careful examination of the Jensen canon that, when it came to the interactions between gene and environment from which the phenotypes developed, Jensen's "remarks clearly demonstrate that he understands neither the mathematical nor the practical problems involved in the estimation of interaction effects." (Chase 1980, 491)

Notably, these appeals to Lewontin and Layzer's arguments have by no means been solely confined to the years of the IQ controversy. In fact, appeals to genotype-environment interaction as a means to undermine heritability measures often continue to be cited as the final word on the debate (see, for more recent examples, Block 1995, Oyama 2000, and Vreeke 2000).[16]

With Lewontin's forceful arguments from the 1970s at hand, why would human behavioral geneticists continue to use heritability measures or attempt to defend Jensen? The quote from Chase (1980) above provides a taste of the answer: Jensen and the subsequent defenders of Jensen were simply ignorant of the force of Lewontin's arguments. Jonathan Kaplan introduced his book *The Limits and Lies of Human Genetic Research* (2000), a sustained criticism on the employment of the analysis of variance and heritability measures by modern behavioral genetics, by noting, "I am only half joking when I say that if more people had read and understood [Lewontin's] seminal work on these issues (much of it done in the 1970's) this book would be unnecessary" (Kaplan 2000, x). So the problem was a simple lack of understanding (see also Block 1995, 110).

But ignorance alone was not enough to account for the persistence of the genetic hypothesis for race differences. Some have speculated that something more sinister was at play. Robert Richardson asked, "How might we explain this blindness [to Lewontin's arguments] on Jensen's part?" (Richardson 1984, 407). Richardson answered, "It is exactly here that the point that his doctrine is a racist doctrine—as it manifestly is—enters in. The latent racism explains the persistence of the view despite its manifest untenability on scientific grounds" (ibid.).

More recently, a minority of scholars have stepped forward to attack the Lewontinian consensus. These Jensenites, though, utilized essentially the same strategy of the Lewontinians when it came to assessing the debate. Neven Sesardic, for example, followed in the footsteps of Jensen by distinguishing a "technical" sense of genotype-environment interaction from a muddle-headed "interactionism" (Sesardic 2005). In Sesardic's terminology, there was a statistical notion of interaction (interaction_S) and a commonsense notion of interaction (interaction_C). And, following Jensen again, Sesardic claimed that much confusion followed from invoking the individual-level, commonsense notion to criticize heritability measures of individual differences, which were only complicated by the population-level statistical notion. "Layzer's argument (defended by many authors) that complexities of developmental processes preclude the possibility of partitioning the phenotypic variation into genetic and environmental components seems to be the result of confusing different levels of analysis" (ibid., 73). Michael Levin, in his defense of *Why Race Matters*, acknowledged that "The most popular reason for discounting genotypic differences, however, is genotype/environment interaction" (Levin 1997, 229). Levin agreed that possible environments might lead to possible interaction. However, Levin followed Jensen in responding, "An obvious objection to this argument is that it very nearly treats what is possible as if it were actual. ... But the sheer *possibility* of environments in which the races agree in (nonzero) intelligence does not show that such environments actually exist, or that, if they do, they could sustain a human society. ... Proponents of the interaction argument offer no evidence that their conjectures are any more than that" (author's emphasis, ibid., 229). So even though Lewontin pointed to the social basis of current developmental environments generating a gap in IQ scores and suggested future developmental environments could eliminate the gap, Levin answered, "In short, even if all 'political' questions look forward to what can be done, gene/environment interaction shows at most that biology *might* be 'neutral,' not that it *is*" (author's emphasis, ibid., 230). Sesardic similarly called Lewontin's focus on norms of reaction "the curious triumph of the *possible* over the *actual*" (author's emphasis, Sesardic 2005, 84–85).

With Jensen's replies to his critics at hand, why would anyone continue to criticize the scientists who measure heritability? Sesardic (2005) followed the Lewontinians in providing an answer, albeit one that was now turned on them. The "mindless cheerleaders" (ibid., 192) suffered from "sheer ignorance" (ibid., 135), employed "egregiously fallacious reasoning" (ibid., 228), and embraced "crude and ill founded" arguments" (ibid., 142).

But, again, ignorance alone was not enough to account for the Lewontinian consensus. Like Jensen, who complained that Layzer's "strong ideological preconceptions" interfered with his analysis, Sesardic also pointed to something more egregious. "The sheer level of ignorance, distortion, and flawed reasoning that characterizes the 'anti-heritability' camp is unprecedented in science and philosophy of science," Sesardic surmised. And then he asked,

> Could it be that, in accordance with the above description [i.e., the weakness of the arguments against heritability], the drastic decline of standards is here due to the dominant intellectual atmosphere, in which those set to undermine heritability can hope to be praised for their political sensitivity and for opposing a dangerous theory, while at the same time they do not have to worry about being severely penalized for possible shortcomings in their logic and methodology? Could this be an explanation? (ibid., 207)

Though rhetorical, Sesardic's answer to the question was fairly obvious.

Explaining the Controversy, Take 2: The Explanatory Divide

There are philosophical and historical reasons we should be dissatisfied with both of the explanations of the IQ controversy offered above.[17] On the philosophical side, the reviews of Lewontin's critique and Jensen's reply reveal that both disputants had a variety of interrelated conceptual, methodological, and empirical arguments for either highlighting or discounting the interaction of nature and nurture. So it is unfair to simply write off one side as ignorant or biased as if that position had no genuinely intellectual foundation. On the historical side, dismissing either the arguments of Lewontin or Jensen as ignorant and politically biased ignores the broader history of the interaction debates by focusing only on the IQ controversy. That is, if we wear historical blinders and only focus on the disputants in this one scientific debate, it is easy to get caught up in the particular personalities and particular cultural context of that one episode. But remember that the arguments of Lewontin were first seen in Hogben 30 years earlier, and the responses of Jensen were found in Fisher's response to Hogben (and, as will be seen in the next chapter, arose again 30 years later). What's more, direct lineages of these arguments can be traced from Hogben to Lewontin and from Fisher to Jensen.

Take Waddington, the developmental geneticist held up by both Lewontin and Layzer as the exemplar of investigating the causal mechanisms of the developmental process. Waddington did most of his work between 1940 and 1970, the years between the eugenics movement and the IQ controversy. Waddington published his major work—*The Strategy of the Gene*—in 1957, over a decade after the eugenics movement eroded and over a decade before the IQ controversy ensued. In *The Strategy of the Gene*, Waddington wanted to explain to his readers what geneticists meant when they spoke of genetic and environmental influences. To do so, he pointed his readers to Hogben's *Nature and Nurture* (1933a), where Hogben introduced interaction as a problem for the analysis of variance (ibid., 94). Like Hogben, Waddington emphasized both the importance of genotype-environment interaction along with the mishandling of it by statistical analyses of variance, arguing, "after nearly a half a century's development the statistical theory still has to leave out of account the contribution of genotype-environment interactions" (ibid., 100).[18]

Contrast Waddington, though, with American agricultural geneticist Jay Lush. Lush, like Waddington, did much of his work in the decades between the eugenics movement and the IQ controversy. But unlike Waddington, Lush was no developmental geneticist; Lush made major contributions to applying the statistical methods of population genetics to animal breeding. And, in fact, Lush is responsible for introducing heritability measures, the very measure Jensen employed so heavily. In Lush's seminal *Animal Breeding Plans* (1937), he brushed aside the importance of genotype-environment interaction in a manner reminiscent of Fisher and prescient of Jensen. "It seems likely," Lush counseled,

> that in general the nonadditive combination of effects of heredity and environment are small in amount* and that many of those which do occur can be reduced to a negligible remainder by choosing a scale of measurement ... which will show the effects of hereditary and environmental variation on that characteristic in their most nearly additive form. (ibid., 64)

The "*" in Lush's statement directed his readers to a footnote where he continued, "For some extreme examples of nonadditive combination effects of heredity and environment consult Chapter 5 of Hogben's *Nature and Nurture*." In contrast to Waddington, then, who introduced Hogben's work as exemplifying what geneticists meant by genetic and environmental influences, Lush relegated Hogben to a footnote, dismissing his work as offering only "extreme examples."[19]

Thus, we should be skeptical of appeals to ignorance and sociopolitical bias as explanations of the trenchant disagreements in the IQ controversy. The debate concerning the interaction of nature and nurture, as reviewed over the last chapter and this one, is larger than the particular disputants involved in any one episode. The disputants on either side have changed, as have the cultural contexts in which those disputants acted. But the arguments on either side have remained remarkably similar. The question, then, remains: how should we make sense of the divide between Lewontin and Jensen?

Interaction: In Obstruction to Variation-Partitioning or in Service of Mechanism-Elucidation?

Let me now offer a different vision of what's going on here, one that takes the interrelated arguments of both sides seriously as well as the broader history of those arguments across the interaction debates. Human behavioral geneticists generally and Jensen specifically, like Fisher before them, took a variation-partitioning approach to studying nature and nurture (refer back to table 2.1 in the previous chapter for a tabular presentation of the components of the variation-partitioning approach). For Jensen, the thing that needed explaining was variation in a population, or, more specifically, the relative contributions of nature and nurture to variation in IQ. Jensen framed his famous 1969 paper in terms of a how-much question: "How Much Can We Boost IQ and Scholastic Achievement?"—a question about how much of the variation in IQ was due to the genetic cause of variation and how much was due to the environmental cause of variation.

A how-much question such as this needed to be answered in terms of the causes of variation responsible for the variation in IQ. And generating that answer involved using a statistical methodology that could partition causes of variation. "The mathematical technique for doing this," Jensen explained, "was invented by Sir Ronald Fisher, the British geneticist and statistician. It is one of the great achievements in the development of statistical methodology" (Jensen 1969, 28).

Lewontin, in contrast, encouraged a mechanism-elucidation approach to studying nature and nurture (refer back to table 2.1 for the components of the mechanism-elucidation approach). "[R]elations between genotype, environment, and phenotype," Lewontin counseled, "are at base mechanical questions" (Feldman and Lewontin 1975, 1167–1168). For Lewontin, like Hogben before him, the thing that needed explaining was the developmental process or, more specifically, the developmental relationship between nature and nurture that played out during the developmental process. Lewontin's mechanical questions were how questions—questions about how differences in genotype and how differences environment interact during development. What did the explaining when it came to these how questions were the causal mechanisms (or "blueprints of this machinery," as Lewontin called it) responsible for the development of a trait.

Jensen and Lewontin, like Fisher and Hogben before them, shared a common interest in studying the relationship between nature and nurture. But, like Fisher and Hogben, Jensen and Lewontin also faced an explanatory divide because they employed very different approaches for that study. It is not surprising, then, that Jensen and Lewontin also reached very different answers to the three questions that separated Fisher and Hogben:

- the conceptual question—what is interaction?
- the investigative question—why and how should interaction be investigated?
- the evidential question—what is the empirical evidence for interaction?

Jensen answered the conceptual question just as Fisher did. For Jensen, interaction was about absence—absence of additivity. It was also, for Jensen, a strictly statistical concept generated by Fisher's analysis of variance. Genotype-environment interaction, for Jensen, was a statistical measure of the breakdown in additivity between the main effects of nature and nurture—Fisher's biometric concept of interaction. Recognizing Jensen's answer to the conceptual question explains why Jensen so quickly dismissed Lewontin and Layzer's emphases on interaction. Jensen treated genotype-environment interaction strictly as a statistical measure descriptive of individual differences in a population, so invoking individual development in a discussion of interaction, Jensen retaliated, arose from a "failure to understand the real meaning of the term 'interaction' as it is used in population genetics" (Jensen 1973, 49).

Jensen also shared Fisher's answer to the investigative question concerning why and how interaction should be investigated. Assessing how much variation in IQ was due to either genes or environment was easiest if the variation in IQ was generated by only genetic and environmental causes of variation. The interaction of nature and nurture posed an obstruction to this effort because it prohibited one from treating differences in IQ simply as the sum of genetic differences and environmental differences. Interaction was to be considered but only in so far as it presented a potential nuisance to getting on with the task of measuring the heritability of IQ. Recognizing Jensen's answer to the investigative question explains why Jensen was undeterred by Lewontin's emphases on possible environments when discussing interaction. When pressed on the locality of the heritability estimates, Jensen simply claimed to confine his genetic hypothesis to the actual environments. As far as Jensen was concerned, Lewontin could speculate all he wanted to about what might happen in as-yet-untested environments, but this speculation would not change what was actually occurring. Abstract speculation about possible environments was a distraction from the task at hand of evaluating the IQ gap in the actual environment, an argument reminiscent of Fisher's distinction between his own "practical" interests and Hogben's "academic" concerns.

When it came to what was actually occurring in actual environments, rather than what might occur in possible environments, Jensen shared Fisher's answer to the evidential question. Interaction was a possible but unproven obstruction. Citing Jinks and Fulker's (1970) survey of twin data available at the time, Jensen claimed there was no evidence of genotype-environment interaction in humans. And so Jensen ultimately fell into line with Fisher by claiming that the burden of proof rested with the critics of the genetic hypothesis to show that interaction was anything more than a possible complication for the heritability estimates of IQ.

Lewontin answered the conceptual, the investigative, and the evidential questions much differently. Remember that Lewontin concluded his 1973 letter to Dobzhansky with the complaint, "Why is it that most geneticists do not understand that the phenotype is a developmental process?" Understanding the phenotype as a developmental process had important implications for how Lewontin conceptualized the relationship between genotype and environment. The genotype, the environment, and the phenotype could not be treated as individual units, according to Lewontin, the first adding to the second to create the third. Rather, the first and the second interacted continuously throughout development, and the third was the manifestation of this interactive, developmental process. Differences in the phenotype, then, would result from differences in this interactive, developmental process. Lewontin pointed to genotype-environment interaction so often because he saw in it a reflection of this developmental relationship. Extreme cases of interaction showed that even slight differences in genetic or environmental distribution could lead to huge differences in phenotypic outcome, thus revealing the interdependence of the factors in this relationship. Lewontin answered the conceptual question just as Hogben did. Genotype-environment interaction was not simply the absence of additivity; it was the presence of something uniquely developmental: variation that resulted from differences in unique, developmental combinations of nature and nurture—Hogben's developmental concept of interaction.

Lewontin's answer to the conceptual question, just as it did for Hogben, shaped his answer to the investigative question about why and how interaction should be sought. Commentators on Lewontin's contribution to the IQ controversy have made much of his passing comment about the analysis of variance being "useless" (Lewontin 1974, 410). But a fixation on that one quote ignores Lewontin's wider writings on the topic. In the 1960 edition of G. G. Simpson's *Quantitative Zoology*, for instance, Lewontin joined Simpson as co-author (Simpson, Roe, and Lewontin 1960). There, Lewontin wrote the chapter on the analysis of variance (Hagen 2003). He warned of its limitations; he warned of the complications posed by the interaction of nature and nurture; he warned of the need to be cautious when making generalizations. But he certainly didn't warn of the analysis of variance being "useless." As Hogben concluded, the danger only arose if it was treated as an end point, rather than as a starting point to other, more interventionist experiments designed to investigate the developmental relationship between nature and nurture. For Lewontin, the next step was studying the norms of reaction generated by exposing different genotypic groups to different environments. Recognizing Lewontin's answer to the investigative question helps explain why he placed so much emphasis on the importance of considering possible environments when discussing these issues. Understanding the developmental process meant understanding the causal mechanisms that responded with different outcomes to different genetic and environmental inputs, and so possible environments were a powerful tool for elucidating such causal mechanisms and testing hypothesized causal models.

When it came to the empirical evidence for genotype-environment interaction, Lewontin pointed his readers to the known cases of the phenomenon in plant and non-human animal studies, as identified by his mentor, Dobzhansky. Gathering such empirical evidence would be much more difficult in humans, Lewontin admitted (Lewontin 1975), but since interaction was developmental in nature it would be the rule and not the exception. Following Hogben, Lewontin placed the burden of proof on those geneticists and psychologists who assumed the additivity of genotype and environment was an accurate reflection of the biological reality.

I have been emphasizing the similarities between the Fisher-Hogben and the Jensen-Lewontin debates over the interaction of nature and nurture. But I don't mean to suggest that the debates were identical. There were important differences, for example, with regard to the academic training and disciplinary identification of these scientists. The Fisher-Hogben debate pitted an evolutionary geneticist against an experimental biologist. Lewontin, by training, was more in line with Fisher than Hogben, and Jensen came to his debate by way of psychology. All of the participants were trained statisticians. Likewise, the domain of the debate shifted from Fisher-Hogben to Jensen-Lewontin. For Fisher and Hogben, the argument was largely over whether interaction existed anywhere in nature (e.g., for eye facets in flies). But for Jensen and Lewontin, it had changed to a dispute over whether interaction existed for IQ in human populations.

The differences between the debates are as instructive as the similarities because they point to what does not explain the divide. Training and disciplinary identification, for example, do not explain the divide. If they did, Lewontin would have been echoing Fisher's arguments rather than Hogben's. Likewise, the presence or absence of statistical expertise does not explain the divide, as all of the disputants in these two chapters (and in the next one) were trained in statistics. What explains the divide, I have been arguing, is a philosophical disagreement about how explanation works in science—about what sort of questions scientists should ask, about what sort of evidence provides answers to those questions, and about what sort of methodologies generate that evidence. It is this explanatory thread that I am tracing because it persists in each new instantiation of the interaction debates despite differences in cast, era, and cultural context.

The Advantage of an Explanatory Divide

Jensen and Lewontin had very different answers to the conceptual, investigative, and evidential questions. But I have tried to show in this chapter that we need not appeal to ignorance or sociopolitical bias to explain those different answers. Answering the conceptual question the way Jensen did doesn't necessarily mean Jensen or his defenders were "racist." Answering the evidential question the way Lewontin did doesn't necessarily mean Lewontin or his defenders were confused or "mindless cheerleaders." Even beyond this historical inaccuracy, such ad hominem attacks slam the brakes on any further rational discourse. How can one side of a debate reasonably respond to being labeled a racist or being dismissed as a mindless cheerleader? It's quite simple. They cannot. And so the conversation reaches a standstill.

An explanatory divide, though, is different. Jensen and Lewontin were largely talking past one another during the IQ controversy because they conceptualized, investigated, and judged evidence of interaction so divergently, and because interaction, for them, was situated in different explanatory frameworks concerning what needed explaining, different appeals to what did the explaining, and different methodologies that generated those explanations. A divide of this nature is subject to further philosophical analysis because it calls out for an intellectual explication of the relationship between those two approaches. In chapters 5 and 6 I will undertake this intellectual explication, attempting to bridge the explanatory divide. But, before proceeding to that bridge, let me bring this history of the interaction debates to the present in chapter 4. For, just as the explanatory divide outlived Fisher and Hogben to arise again in the IQ controversy, so too did it outlive the IQ controversy. It has emerged again in the twenty-first century.

4 Its Rise, Its Fall, Its Rise? Interaction in the Twenty-First Century

In the May 2011 issue of *Archives of General Psychiatry*, the editors published a massive meta-analysis. A meta-analysis combines the data and results from previous studies and then analyzes the set of studies as one large data set. This meta-analysis combined 54 previous studies that all examined the relationship between the serotonin transporter gene, exposure to various stressful life events, and the subsequent development of depression. With 54 studies included, the meta-analysis combined data from over 40,000 human research participants. The meta-analysis designers concluded from their study that the serotonin transporter gene moderates the effect of stress exposure when it comes to depression—that the serotonin transporter gene and stress provide a case of gene-environment interaction (Karg et al. 2011).

When the results of the meta-analysis were reported in the media, however, something peculiar emerged. The *Los Angeles Times* mocked, "Ah, the process of science. One week, oat bran is the font of eternal youth. The next week, it kills. It's not clear what these new results mean. Or what next week might bring" (Mestel 2011). Even more telling, the editors of *Archives* included a commentary in the same issue as the meta-analysis, and there John Hardy and Nancy Low worried, "The reader is therefore entitled to ask, 'What should I believe? Which explanation is true?' Unfortunately, the answer is unclear, and a long time will pass before questions can be resolved because all the studies so far can be interpreted in opposing ways" (Hardy and Low 2011, 455). Hardy and Low's reference to "all the studies so far" pointed to the fact that the 54-study meta-analysis in *Archives* was not the first meta-analysis of studies on the serotonin transporter gene, stress, and depression; in fact, it was the third. And the previous two came to the exact opposite conclusion—no gene-environment interaction. The *Los Angeles Times* article was subtitled "Its Rise, Its Fall, Its Rise?", an allusion to the dilemma. There was initially much hype about this case of gene-environment interaction, but then the early meta-analyses questioned that initial result ... and then the most recent meta-analysis seemed to do the exact opposite.

How can meta-analyses of the same phenomenon generate different results? And, when they do, "What should I believe? Which explanation is true?" The purpose of this chapter is to answer these questions. My thesis is that the explanatory divide that separated Fisher and Hogben in chapter 2 and then Jensen and Lewontin in chapter 3 also separated participants in this most recent debate over gene-environment interaction. The chapter will bring the history of research on interaction to the present and, in so doing, show how the explanatory divide that plagued past debates continues to persist into the present day.

Its Rise: "A Watershed Event in the History of Behavioral Genetics"

On April 1, 1972, researchers in Dunedin, New Zealand, began keeping track of every baby born in the city's Queen Mary Hospital. This continued for one full year, until March 31, 1973, during which just over 1,000 children were born. These babies were now enrolled in the Dunedin Multidisciplinary Health and Development Study (or just "the Dunedin Study"). The Dunedin Study, brainchild of educational psychologist Phil Silva from the University of Otago, was originally designed to track these children for just a few years in order to examine whether complications associated with pregnancy and birth led to health and developmental problems in early childhood, and so the researchers met back up with the participating children and their parents three years later. The study, however, did not end there. The researchers quickly recognized that they had a unique resource in the Dunedin Study—a representative sample of New Zealand children identified and willing to participate (initially through their parents' consent) in long-term observation. A series of national and then international grants allowed the researchers to meet back up with the children at ages (or "phases") 5, 7, 9, 11, 13, 15, 18, 21, 26, 32, and 38. The participants are now in their 40s, with the next scheduled meetings to occur at phases 44 (2016–2017) and 50 (2022–2023). Over the years, the researchers have gathered a wealth of information about the study participants' environments (e.g., breastfed or not, exposed to child abuse or not) and the study participants' resulting traits (e.g., IQ scores, criminal records). In 1998–1999, during phase 26, the researchers obtained DNA samples from the study participants with their consent, adding genetic information to the available data (Moffitt, Caspi, Rutter, and Silva 2001).

Part of the problem with the eugenics and IQ controversies discussed in the last two chapters was simply that there were so little human data upon which to base empirical claims. With plant and non-human animal studies, geneticists could raise genetically different groups of potatoes or fruit flies in various environments in order to experimentally examine how the different genetic groups responded to the different environments. Hogben, in the 1930s, and then Lewontin, in the 1970s, drew attention to such studies in an effort to point out that interaction was common in nature. But if a geneticist wanted to examine the same phenomenon in humans, she had to rely on rare populations of twins and adoptees in order to see how different genetic groups responded to different environments. The ethics of research obviously prevented the geneticist from, say, separating identical twins at birth just to see how the same genome responded to two different environments. Francis Galton, as mentioned in chapter 1, recognized twins as nature's experiment, but nature does not provide the geneticist with an endless supply of such experiments. As a result, disputants from the eugenics controversy and then the IQ controversy were forced to make assumptions about how common (if at all) interaction was in humans.

That problem goes away, however, if the geneticist switches from investigating entire genomes to investigating particular genes. Suppose a geneticist is interested in investigating how nature and nurture interact to bring about the risk of developing clinical depression. Nurture, in this case, is exposure to stressful life events (such as the loss of a loved one, childhood maltreatment, or a major illness). If, on the nature side, the geneticist is confined to entire genomes, then she must find sets of twins who have been exposed to different numbers of stressful life events (as human behavioral geneticists did throughout the twentieth century). So now the rare population of twins is made even smaller by the constraints of the environment being studied. But if, on the nature side, the geneticist looks at a particular gene every human has, then every human is a potential research participant. That is, if this common gene has different variants, then individuals with one variant can be grouped together as a genetic group and individuals with another variant can be grouped together as a genetic group, and the geneticist can examine how clinical depression arises in the two different groups depending on how many stressful life events are experienced. By switching from entire genomes to particular genes, everyone is now a potential participant in nature's experiment. Participants in the Dunedin Study, by providing their environmental and then their genetic information, offered researchers a unique opportunity to study nature's experiment.

Collecting DNA from the Dunedin Study participants was the initiative of spouses Terrie Moffitt and Avshalom Caspi, who were psychologists at the University of Wisconsin at the time.[1] In a 1994 keynote address to the International Society for the Study of Behavioral Development, human behavioral geneticist Robert Plomin encouraged psychologists to collect DNA from participants in the studies that they were running because "the genetics of behavior was too important to be left to geneticists." Moffitt and Caspi were already affiliated with the Dunedin Study; Moffitt joined the New Zealand team in 1984, acting as Silva's adjunct deputy director of deviance research and assisting with the phase 13 data collection. Inspired by Plomin's recommendation, Moffitt and Caspi secured research funds from the University of Wisconsin to include DNA extraction in the phase 26 data collection.

Around this same time, Moffitt and Caspi were invited by Sir Michael Rutter to join his newly formed Social, Genetic, and Developmental Psychiatry Centre (SGDP) at the Institute of Psychiatry, affiliated with King's College London. Rutter, called the "father of British child psychiatry," set up the SGDP in 1994 two years after he was knighted with the hopes of creating an interdisciplinary unit devoted to bringing diverse viewpoints to bear on understanding the development of and differences in psychiatric diseases and disorders (McGuffin and Plomin 2004). Rutter had already convinced Plomin to join the team (as well as behavioral geneticist Thalia Eley). Molecular geneticists Ian Craig and David Collier, psychiatrists Philip Asherson, David Ball, and Pak Sham, and developmental psychologists Judy Dunn and Francesca Happé were also there. For Rutter, the team of Moffitt and Caspi offered him a pair of psychologists on the rise with access to the extraordinary Dunedin Study; for Moffitt and Caspi, the SGDP offered them a unique opportunity to work with an interdisciplinary group of scientists trained in methodologies such as twin studies and molecular genetics that they didn't learn as part of their psychological training. In 1997, they left Wisconsin (but kept their university affiliation) and moved to London.

When Moffitt and Caspi arrived at the SGDP in the late 1990s, geneticists had already identified a number of candidate genes thought to be associated with complex human traits. In 1996, the dopamine D4 receptor gene (*DRD4*) was linked up to novelty seeking, and variations in the promoter region of the serotonin transporter gene (*5-HTTLPR*) were connected with depression (Benjamin et al. 1996; Ebstein et al. 1996; Lesch et al. 1996). Other candidate genes included monoamine oxidase A (*MAOA*) and catechol-O-methyltransferase (*COMT*). (I will explain the molecular and functional biology of these genes below.) Within just a few years, though, the initial excitement about these candidate genes was tempered by follow-up studies that did not find the same gene-behavior associations. Why? An initial thought at the time was that the inability to reliably replicate the original findings was the result of measurement error—that scientists simply needed to refine the phenotype with better measures of it. Moffitt and Caspi believed that, if measurement error was the problem, then the Dunedin Study was the solution; the remarkable breadth and depth of behavioral measures of that study population ensured that the phenotype was about as refined as it could be. With the newly acquired DNA information about the research participants, the SGDP team set off to sequence the most promising genes and check them against the carefully measured phenotypic information from Dunedin.

Moffitt and Caspi prepared for a series of association studies with the Dunedin Study: *DRD4* was to be linked up with attention deficit and hyperactivity disorder; *MAOA* was to be connected with antisocial behavior; the serotonin transporter gene was to be associated with depression. But when the results of the first study came in, rather than launching Moffitt and Caspi's new gene-behavior research approach, it brought it to a screeching halt. Their 2002 publication on *DRD4* concluded, "In summary, data from this birth cohort that have contributed to the ADHD literature for some time, provide little evidence to support the association of the seven-repeat allele of *DRD4* with quantitative measures of hyperactivity from ages 7 to 15; nor with the presence of clinically-defined ADHD; nor with personality trait measures of impulsivity in adolescence or adulthood" (Mill et al. 2002, 389). It was their "*DRD4* experience of nothingness," and it convinced Moffitt and Caspi that the problem with trying to link up candidate genes with complex human traits was not just an issue of measuring the phenotype better. There was a more fundamental problem with the gene-behavior approach. So rather than risking repeated failures, they took a step back and rethought the guiding principle that would direct their research.

That step back coincided with a trip that Moffitt and Caspi took to Africa. There they saw how people lived in locales with high infestation of mosquitoes and in communities with high rates of HIV—known environmental causes of malaria and AIDS. But even though everyone was exposed to mosquitoes, not everyone developed malaria, and even though many people were exposed to HIV, not all of them developed AIDS. The existing idea was that something biological protected certain individuals from these environmental pathogens. Moffitt and Caspi thought a similar host-pathogen model might make for a guiding principle to studying complex human traits like antisocial behavior and depression. *MAOA*, rather than being treated in isolation, would be combined with a known environmental "pathogen" of antisocial behavior: childhood maltreatment (such as physical abuse, sexual abuse, or neglect). The serotonin transporter gene, rather than being treated in isolation, would be combined with a known environmental "pathogen" of depression—stress. The Dunedin Study continued to serve Moffitt and Caspi, but now it was the extensive and carefully measured environmental information it contained about the research participants that proved so valuable.

With this research model in place, Moffitt, Caspi, along with their collaborators in Dunedin and London, published a series of articles reporting significant interactions between particular genes and particular environments for a range of human traits. The first, published in *Science* in 2002, examined the relationship between high versus low levels of monoamine oxidase A produced by different variants of the *MAOA* gene, exposure to childhood maltreatment, and the development of antisocial behaviors such as criminal violence (Caspi et al. 2002). MAOA is a metabolic neuroenzyme that inactivates neurotransmitters such as dopamine and serotonin. Antisocial personality disorder is a clinical disorder defined by a predatory pattern of manipulating or violating the rights of others; not surprisingly, it is correlated with criminal violence. As can be seen in figure 4.1, Moffitt and Caspi found that low-MAOA individuals were more likely to be convicted of criminal violence when exposed to childhood maltreatment than were high-MAOA individuals.[2]

In their second study published the following year (again in *Science*), Moffitt and Caspi reported on the relationship between short versus long forms of the serotonin transporter gene (*5-HTTLPR*), exposure to stressful life events, and the development of clinical depression (Caspi et al. 2003). Serotonin transporters are proteins on the synaptic membranes of neurons involved in the reuptake of neurotransmitters after they have been released into the synapse. The efficiency of these proteins produced at a synapse is regulated by the promoter region of the serotonin transporter gene, which is itself based on the length of that promoter region. Individuals receive one copy from each of their parents, which can be either short (s) or long (l); thus, individuals can be classified as either s/s, s/l, or l/l. As can be seen in the norm of reaction graph in figure 4.2, Moffitt and Caspi found individuals with the short form of the serotonin transporter gene to be more vulnerable to developing depression when exposed to repeated stressful life events than were individuals with the long form of the gene.

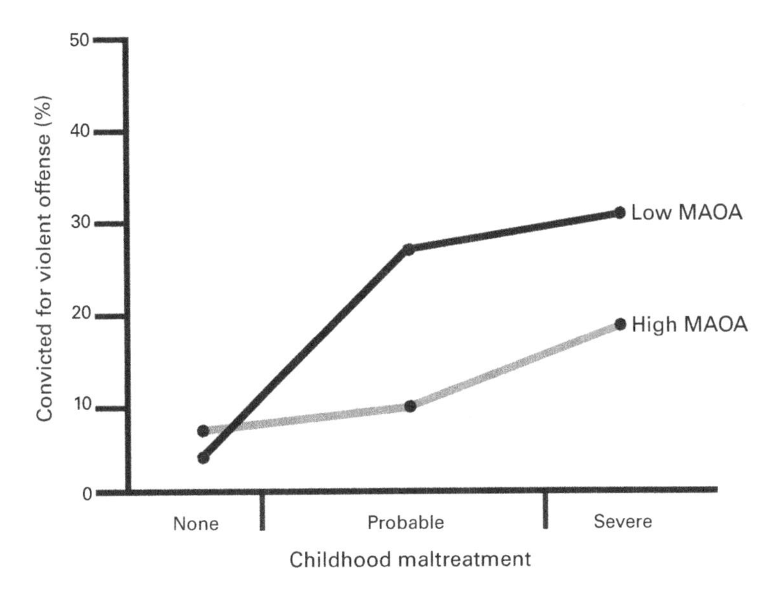

Figure 4.1
Norm of reaction graph for monoamine oxidase A (*MAOA*), childhood maltreatment, and convictions for violent offenses (adapted from Caspi et al. 2002, figure 2B).

In their third study, Moffitt and Caspi examined the relationship between allelic variants in the catechol-O-methyltransferase gene (*COMT*), adolescent use of cannabis, and the development of psychotic symptoms associated with schizophrenia (Caspi et al. 2005). The *COMT* gene produces catechol-O-methyltransferase, which is involved in the metabolism of dopamine released into synapses. The gene comes in two forms based on whether the amino acid produced by the DNA is valine (Val) or methionine (Met); thus the three possibilities are Val/Val, Val/Met, and Met/Met. The Met version of the gene leads to less catechol-O-methyltransferase activity, resulting in lower metabolism of dopamine in brain regions also affected by cannabis. As can be seen in figure 4.3, Moffitt and Caspi found individuals with either the Val/Met or, especially, the Val/Val forms were much more vulnerable to developing psychosis if cannabis was used during adolescence.

Finally, in a 2007 study, Moffitt and Caspi investigated the relationship between variants in the fatty acid desaturase 2 gene (*FADS2*), exposure to breastfeeding, and IQ (Caspi et al. 2007). The *FADS2* gene produces fatty acid desaturase 2, which is an enzyme responsible for breaking down dietary fatty acids as found in, for instance, breast milk. Individuals carry

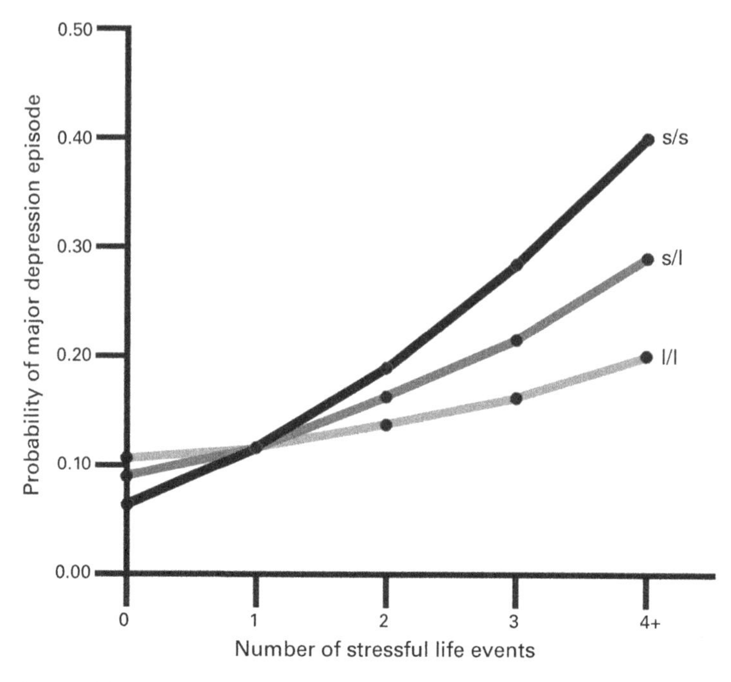

Figure 4.2
Norm of reaction graph for serotonin transporter gene (*5-HTTLPR*), stress, and depression (reprinted from Caspi et al. 2003, figure 1B, with permission from AAAS).

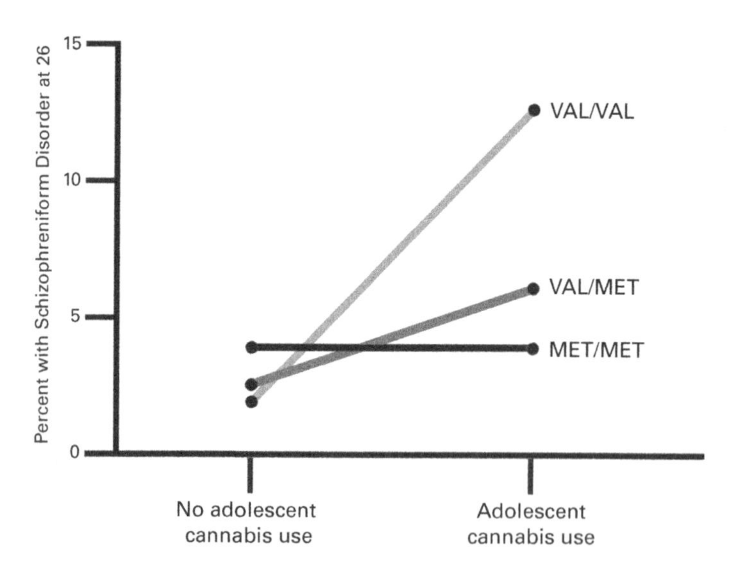

Figure 4.3
Norm of reaction graph for catechol-O-methyltransferase gene (*COMT*), adolescent cannabis use, and psychosis (adapted from Caspi et al. 2005, figure 1A).

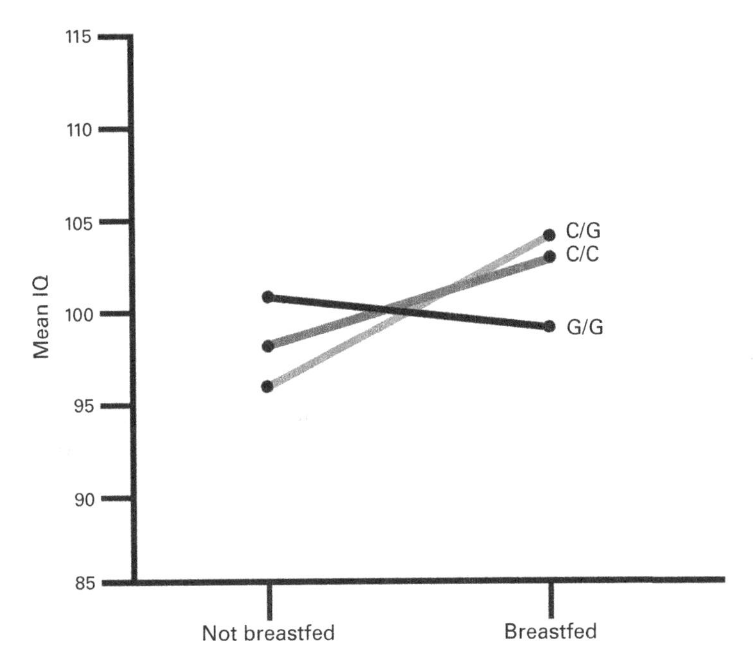

Figure 4.4
Norm of reaction graph for fatty acid desaturase 2 gene (*FADS2*), breastfeeding, and IQ (adapted from Caspi et al. 2007, figure 1A).

either a cytosine (C) or guanine (G) within this gene; thus the three possibilities are C/C, C/G, and G/G. In this instance, Moffitt and Caspi used environmental exposure to dictate which gene to study in the gene-environment relationship, as *FADS2* had not previously been considered in relation to intelligence. As can be seen in figure 4.4, Moffitt and Caspi found that individuals with either the C/C or C/G forms who were breastfed had significantly higher IQs on average than individuals with either the C/C or C/G forms who were not breastfed. Breastfeeding did not significantly affect IQ scores for individuals with the G/G form.

It is hard to overstate the reception of these four publications. In contrast to scientific studies that investigated only genetic contributions or only environmental contributions to complex human traits, Moffitt and Caspi combined these, dealing an apparent deathblow to "nature versus nurture." The 2002 study on *MAOA*, childhood maltreatment, and antisocial behavior was hailed in the *Economist* (2002) and *Popular Mechanics* (Wilson 2002). The 2003 study on the serotonin transporter gene, stress, and depression was covered in hundreds of print, television, and radio stories worldwide and even featured in a lengthy article in *The New York Times Magazine* (Bazelon 2006). The 2005 paper on the *COMT* gene, adolescent cannabis use, and psychosis was highlighted in a 2007 report from the Executive Office of the U.S. President on the relationship between marijuana and mental illness (Office of National Drug Control Policy 2007). And the 2007 study on *FADS2*, breastfeeding, and IQ was reported in the *Los Angeles Times* (Kaplan 2007), *Scientific American* (Swaminathan 2007), and *Newsweek* (Begley 2008).

Importantly, this excitement was not confined to the popular press. The editors at *Science* selected the serotonin transporter gene study as runner-up for "breakthrough of the year" in 2003 (News and Editorial Staffs 2003).[3] Behavioral geneticist Dean Hamer, in a review article for *Science*, identified the *MAOA* study as paving the way for the future of human behavioral genetics (Hamer 2002). Legal scholar Robert Stone forecasted, "the Caspi Study may prove to be the beginning of a new era" (Stone 2003, 1559). And bioethicist Erik Parens wrote, "It might not be an exaggeration to say that, if replicated, the Caspi-Moffitt MAOA study will turn out to have been a watershed event in the history of behavioral genetics" (Parens 2004, S22). As of summer 2013, the *MAOA* study has been cited nearly 3,000 times, the serotonin transporter gene study over 5,000 times, and the *COMT* and *FADS2* studies both several hundred times.[4]

Its Fall: "A Field That Was Eager for a New Approach"

Parens claimed that the Moffitt-Caspi research would be a watershed event *if* the results were replicated. In science generally, and genetics especially, independent replications of an original result are absolutely necessary to confirm or disconfirm the purported result. An independent replication occurs when scientists at another laboratory and studying another population investigate whether or not the original result arises. The independent replication is performed at another laboratory in order to ensure that the original researchers' biases (be they conscious or unconscious) do not affect the follow-up investigation. And the independent replication is performed with another population in order to determine whether or not the original result is generalizable or simply an idiosyncrasy of the original population.

The high level of attention and excitement surrounding Moffitt and Caspi's four studies on gene-environment interaction made them natural candidates for independent replication, and so scientists at other laboratories set out to investigate whether they would find the same results concerning antisocial personality disorder, depression, psychosis, and IQ in different populations. The study that received the most attention was the one on depression.[5] Within six years of the original publication, dozens of independent replications were performed and published, involving populations ranging from Spain to Japan and Greece to China (e.g., Cervilla et al. 2007; Katsuyama et al. 2008; Veletza et al. 2009; Zhang et al. 2009).[6] Some of these independent replications of the depression study were positive or partially positive replications, where a partially positive replication finds the result in, for example, females but not males (for a selection, see Eley et al. 2004; Kaufman et al. 2004; Kendler et al. 2005; Grabe et al. 2005; Nakatani et al. 2005; Sjöberg et al. 2006; Taylor et al. 2006; Jacobs et al. 2006; Kaufman et al. 2006; Wilhelm et al. 2006; Zalsman et al. 2006; Cervilla et al. 2007; Kim et al. 2007; Mandelli et al. 2007; Brummett et al. 2008; Lazary et al. 2008; Katsuyama et al. 2008; Aguilera et al. 2009; Veletza et al. 2009). Some of the independent replications of the depression study were negative replications, also called "failures to replicate" or "nonreplications" (see, for example, Gillespie et al. 2005; Surtees et al. 2006; Chipman et al. 2007; Covault et al. 2007; Middeldorp et al. 2007; Scheid et al. 2007; Araya et al. 2009). And some of the independent replications of the depression study were opposite of the original, discovering the long form of the serotonin transporter gene to be more vulnerable to stress exposure rather than the short form (Sjöberg et al. 2006; Chorbov et al. 2007; Laucht et al. 2009; Veletza et al. 2009; Zhang et al. 2009).

This mixed bag of positive and negative replications is not uncommon in genetics, and it does present a puzzle. If independent replications of an original study are supposed to lend support for or against the original result, what does a combination of positive and negative replications say about the original result? Is it supported or discredited? To solve this problem, scientists commonly turn to meta-analyses. A meta-analysis, as explained in the introduction to this chapter, takes data from multiple studies and combines them all into one large data set. This large data set is then subject to analysis in order to determine whether the original result arises here. If it does, then the meta-analysis is taken to lend support to the original result; if it does not, then the meta-analysis is taken to be evidence against the original result.

In 2009, Kathleen Ries Merikangas of the National Institute of Mental Health (NIMH) led a meta-analysis on independent replications of the serotonin transporter gene, stress, and depression (Risch et al. 2009). The meta-analysis followed on the heels of a 2006 workshop at the NIMH in which the goal was to assess the data on gene-environment interaction (ibid., 2463). The mixed bag of positive and negative replications, however, meant that there was plenty of room for disagreement. "Because there was a lack of consensus on the status of replication of gene-environment interaction studies," the authors explained, "a comprehensive review and meta-analysis of this research was proposed" (ibid.). Of the dozens of independent replications available at that time, the team included 14 for meta-analysis (a point I will return to below): Eley et al. (2004), Gillespie et al. (2005), Grabe et al. (2005), Surtees et al. (2006), Taylor et al. (2006), Wilhelm et al. (2006), Kim et al. (2007), Cervilla et al. (2007), Chipman et al. (2007), Chorbov et al. (2007), Middeldorp et al. (2007, 2008), Laucht et al. (2009), and Power et al. (2010). The studies they combined had anywhere from 118 to 4,175 participants, creating a meta-analysis with over 14,000 total participants analyzed.

Merikangas's study was a resoundingly negative meta-analysis of Moffitt and Caspi's original gene-environment interaction result. They did find a significant main effect for stressful life events on depression. However, their meta-analysis found no significant main effect for the serotonin transporter gene and no significant interaction between the serotonin transporter gene and stress. The authors concluded their paper with a criticism of "a field that was eager for a new approach" (Risch et al. 2009, 2469).

In part due to the intense attention given to Moffitt and Caspi's original gene-environment interaction results, the negative meta-analysis by Merikangas and her team generated its own media stir. *The New York Times* reported that the original research by Moffitt and Caspi was "now faulted" and had "not held up to scientific scrutiny" (Carey 2009). *The Wall Street Journal* declared the original research "not valid" (Wang 2009).[7]

Merikangas's meta-analysis received the most media attention, but it was not the first meta-analysis published on the relationship between the serotonin transporter gene, stress, and depression. Earlier that same year, a British team led by Marcus Munafò published their own meta-analysis (Munafò et al. 2009). Of the dozens of independent replications published at that time, this team included five for meta-analysis (again, a point I will return to below): the original Caspi et al. 2003 study, Eley et al. 2004, Surtees et al. 2006, Kim et al. 2007, and Scheid et al. 2007. And like the Merikangas meta-analysis, Munafò and his colleagues reported a negative result. They also found a significant main effect for stressful life events on depression, but they also found no significant main effect for the serotonin transporter gene and no significant interaction.

An Alternative Approach: Genome-Wide Association Studies

Merikangas and Munafò, in addition to sharing authorship of negative meta-analyses, also shared a vision for what geneticists ought to be studying. The opening sentence of Merikangas's meta-analysis pointed to it: "The successful statistical identification and independent replication of numerous genetic markers in association studies have confirmed the utility of the genome-wide approach for the detection of genetic markers for complex disorders" (Risch et al. 2009, 2462). This was a reference to genome-wide association studies. In a genome-wide association study, researchers create two large groups of human research participants (often including several thousand participants in each). One group (the case group) consists of participants who all share a common disease or disorder, while the other group (the control group) consists of participants who all lack that common disease or disorder. The researchers then look at the whole genomes of all the participants in both groups, looking for any genetic variants that are more common in the case group than the control group. Genome-wide association studies have now been performed on dozens of human traits, ranging from medical diseases like Crohn's disease and multiple sclerosis to psychiatric disorders such as schizophrenia and bipolar disorder and to morphological characteristics such as bone mineral density and height. The studies have turned up thousands of new genetic associations.[8]

Genome-wide association studies and gene-environment interaction studies actually share something of a common origin. They both emerged in response to the failure of the candidate gene approach to studying complex human traits. Remember, when Moffitt and Caspi failed to link up a candidate gene (*DRD4*) with attention deficit hyperactivity disorder, they decided to instead study future candidate genes in combination with environmental causes of traits. Genome-wide association studies, rather than combining a gene with an environmental cause, instead searched for many genes all contributing to the trait in question. Merikangas and Munafò opted for this second approach. Merikangas and her meta-analysis co-author Risch, for example, are often credited with first making the case for the genome-wide association study approach (Forero 2009). In a short perspective paper for *Science* published almost a decade before the first genome-wide association study, they imagined a future where thousands of single-nucleotide polymorphisms were identified and then associated with all sorts of human traits in studies involving thousands of people (Risch and Merikangas 1996). The great number and variety of genome-wide association studies that have been published since 2005 suggests their call was heard. Thus their conclusion in a 2009 interview about that prophetic paper: "It appears that time has borne out the logic of our conclusions" (Forero 2009). And hence the opening sentence of their negative meta-analysis, which praised the confirmed utility of the genome-wide association study approach to studying complex human traits. Munafò as well has been active in the genome-wide association study approach. He has participated in empirical research (Tozzi et al. 2013), and he has made the case for opting for the genome-wide association study approach over the gene-environment interaction approach (Flint and Munafò 2013). I will return to the relationship between these two approaches below.

Its Rise? "This Is the Final Word"

Two years after Merikangas and Munafò's meta-analyses, Srijan Sen, from the University of Michigan, led the third meta-analysis, the one that opened this chapter. This was not the first time that Sen investigated the serotonin transporter gene *5-HTTLPR*. In 2004, he led a meta-analysis that looked at studies examining the association between the serotonin transporter gene and anxiety-related personality traits like depression (but without any environmental variable in the analysis); he reported a significant association between *5-HTTLPR* and neuroticism, a non-significant association between *5-HTTLPR* and harm avoidance, and no association between *5-HTTLPR* and other anxiety-related personality traits (Sen, Burmeister, and Ghosh 2004). Several years later, Sen added an environmental stressor into the mix—medical residency (the exhausting and highly stressful hospital internship following medical school). Sen tracked 740 residents at 13 hospitals and found residents who carried the short form of *5-HTTLPR* were at significantly greater risk of depressive symptoms during residency (Sen et al. 2010).

Turning to the 2011 meta-analysis: in contrast to Merikangas's meta-analysis with 14 studies and Munafò's meta-analysis with five studies, Sen's 2011 meta-analysis included 54 studies. In contrast to Merikangas's meta-analysis with over 14,000 participants and Munafò's meta-analysis with over 6,000 participants, Sen's meta-analysis combined data from over 40,000 human research participants. And, most importantly, in contrast to Merikangas's and Munafò's discrediting negative results, Sen and his colleges obtained a positive result supportive of Moffitt and Caspi's original finding. The serotonin transporter, in Sen's meta-analysis, moderated the effect of exposure to stress and, in particular, specific stressors like childhood maltreatment or a specific medical condition when it came to depression (Karg et al. 2011).

The advocates for research on gene-environment interaction were quick to hail Sen's meta-analysis over the previous ones. Moffitt, for instance, told a reporter covering the story that Sen and his colleagues' "careful and systematic approach reveals why the 2009 meta-analysis got it wrong. ... We hope that the same journalists who were so hasty to publish a simplistic claim in 2009 will cover this more thoughtful new analysis" (University of Michigan Health System 2011). Sen told WebMD, "This is the final word" (Mann 2011).

But, as I introduced at the beginning of this chapter, not all commentators were so convinced by this "final word" summation. Rosie Mestel at the *Los Angeles Times* argued, "It's not clear what these new results mean. Or what next week might bring" (Mestel 2011). Precisely because of this, other science reporters became hesitant to even cover the latest twist. When Moffitt brought Sen's meta-analysis to the attention of Benedict Carey, who covered the Merikangas meta-analysis for *The New York Times* in 2009, Carey replied, "Dueling meta-analyses do not make for clear science or science reporting."[9]

The problem can be summarized as follows: after Moffitt and Caspi published their original study on the interaction between the serotonin transporter gene and stress, a series of independent replications were performed to see whether that original result held up. But the independent replications presented a mixed bag: some were positive and supported the original result, while some were negative and discredited the original result. When presented with this mixed bag, scientists took the natural next step and performed meta-analyses. But rather than solving the problem, the meta-analyses only perpetuated it by presenting their own mixed bag. The meta-analyses were supposed to provide the meta-solution, but instead they only elevated it to a meta-problem. Hardy and Low (2011), in their commentary on the Sen meta-analysis, asked the obvious questions: what should you believe when dueling meta-analyses reach opposite conclusions? Which explanation of the phenomenon is true?

"Dueling Meta-Analyses"

Meta-analyses, because they combine data from many separate studies, are held in the highest regard as the best form of evidence for or against a purported phenomenon. But Jacob Stegenga has cautioned that it's not always that simple (Stegenga 2011). A mixed bag of positive and negative meta-analyses, Stegenga points out, is actually quite common. The reason is that meta-analysis investigators encounter a number of decision points in the design and implementation of meta-analyses, and there are a variety of reasons why one meta-analysis designer will opt one way at a decision point, while another meta-analysis designer will opt another way. One such decision point is the choice about which studies to include in the meta-analysis and which studies to exclude. Although there have been dozens of studies examining the serotonin transporter gene, stress exposure, and depression after the original Moffitt-Caspi study, these studies have not all been identical, in part because the phenomenon under investigation is so complex. There are a variety of different ways to measure depression, and so meta-analysis designers must decide how similar each study's measure of depression needs to be to the original study in order to justify including it. Likewise, there are a variety of different experiences that can count as a stressor, and so meta-analysis designers must decide how similar each study's measure of stress needs to be to the original study in order to justify including it.

Merikangas's team, Munafò's team, and Sen's team each had to decide which studies to include and which to exclude from their meta-analysis. Merikangas and Munafò, on the one hand, and Sen on the other made radically different decisions. Munafò's meta-analysis included five studies, while Merikangas's meta-analysis included 14 studies, and Sen's meta-analysis included 54 studies. Now, you might think the difference here is simply a discrepancy in date: Sen's meta-analysis was published in 2011, two years after Merikangas's and Munafò's meta-analyses; and so, the thought goes, Sen simply had access to more studies. But that's not what is going on. Sen included only replications published up until December 2009, only months later than Merikangas's cutoff date (March 2009); Munafò's cutoff date was earlier (December 2007), but even then there were dozens of replications to choose from. So it was not just a question of timing. No, Merikangas, Munafò, and Sen all had dozens of potential studies to choose from; Merikangas and Munafò opted for a *restrictive* approach to study inclusion, while Sen opted for a *permissive* approach to study inclusion. The question then becomes: why choose to be either restrictive or permissive when it comes to deciding which studies to include in a meta-analysis?

Let's start with the restrictive meta-analyses run by Merikangas and Munafò. To be included in these meta-analyses, replications needed to match almost identically Moffitt and Caspi's original environmental measure of stress. Moffitt and Caspi, recall, obtained information from the

Dunedin Study participants about their exposure to stressful life events, such as the death of a loved one or the loss of employment. They then grouped the participants based upon how many stressful life events they had encountered: 0, 1, 2, 3, or 4+ (see figure 4.2 above). Merikangas's and Munafò's teams only included replications that similarly combined many different stressors together as "stressful life events," wherein the participants could be grouped across studies based on how many of these stressful life events they had encountered. The meta-analyses by Merikangas and Munafò were thus testing the legitimacy of Moffitt and Caspi's original finding strictly interpreted: same gene, same environmental measure, same measure of depression. Under that strict interpretation, the relatively few studies included generated a negative result.

Boldly, Merikangas's team also suggested they had tested the legitimacy of the entire Moffitt-Caspi research approach to studying gene-environment interaction. Although their meta-analysis only focused on the depression study, they warned that Moffitt and Caspi's other studies on antisocial personality disorder, psychosis, and IQ were also now suspect (Risch et al. 2009, 2469). And so they concluded, "Despite the lack of valid confirmation of the Caspi et al. [2003] results, the approach to implicate candidate genes that have failed previous direct association studies through inclusion of an environmental exposure has been rapidly embraced, and substantial resources have been devoted to subsequent research" (ibid.). The meta-analysis, then, was not to be interpreted as just a refutation of one particular finding; it was to be interpreted as an indictment of an entire "field that was eager for a new approach" and that had already devoted "substantial resources" to this questionable approach (ibid.).

Contrast the restrictive meta-analyses by Merikangas's and Munafò's teams with the permissive meta-analysis by Sen and his team. Sen recognized that a number of studies following Moffitt and Caspi's original result did not attempt to strictly replicate that original study but rather attempted to explore whether the original result would appear under slightly different environmental measures of stress. A number of replications, rather than grouping a number of different environmental stressors together as "stressful life events," focused on a specific stressor in order to examine how individuals with different forms of the serotonin transporter gene responded to that specific stressor. These replications investigated medical conditions like exposure or lack of exposure to a heart attack (Nakatani et al. 2005), hip fracture (Lenze et al. 2005), and stroke (Ramasubbu et al. 2006); they also investigated exposure or lack of exposure to non-medical stressors like experiencing a hurricane (Kilpatrick et al. 2007).

Notice the dates on those studies; they all pre-date the cutoff for study inclusion in the Merikangas and Munafò meta-analyses. Those studies were available to those earlier meta-analysis designers, but they chose not to include them. They didn't include them because exposure or lack of exposure to a single, specific stressor (like a heart attack or a hurricane) is not the same thing as exposure to various numbers of various different stressors.

When Moffitt and Caspi replied to journalists covering the Merikangas meta-analysis and wrote to the journal that published the Munafò meta-analysis, they emphasized how the restrictive approach taken by those two meta-analyses misrepresented the complete body of scientific evidence. The Merikangas meta-analysis "not only opted not to analyse, but also failed to mention, the existence of well-designed studies that compared individuals exposed to stress" such as the stroke and hip fracture studies (Moffitt and Caspi 2009).[10] Moreover, Moffitt and Caspi argued, the meta-analysis designers failed to mention the existence of experimental studies on humans and non-human animals. By 2009, neuroscientists like Daniel Weinberger and Ahmad Hariri (who I will come back to in the next two chapters) had designed interventionist experiments to examine how humans with different forms of the serotonin transporter gene responded to stress exposure and reported that individuals with the short form showed greater stress response on measures of brain circuitry (Hariri et al. 2005; see also Beevers et al. 2007 on a measure of cognitive reactivity). Likewise, other scientists designed stress-inducing experiments on mice and monkeys and similarly reported higher levels of stress response among the animals with the short form analog of the serotonin transporter gene (Champoux et al. 2002; Barr et al. 2004; Carola et al. 2008).

Moffitt and Caspi concluded their reply, "Science aside, we are musing about why these respected scientists are concerned that, as they write 'substantial resources have been devoted to' research on genetic sensitivity to environmental causes of disease" (Moffitt and Caspi 2009). Funding for research on gene-environment interaction, they countered, was miniscule in comparison to research on genome-wide association studies. "What is needed is not less research into gene-environment interaction," they rebuked the Merikangas team's meta-analysis, "but more research of better quality, and more thorough and thoughtful evaluation of it" (ibid.). And in a separate reply to the Munafò team's meta-analysis, they countered, "we believe more work is needed to unravel the mechanisms by which stress might confer vulnerability to depression in individuals with the 's' allele of *5-HTTLPR*" (Kaufman et al. 2010, e19).

These criticisms of the negative meta-analyses did not go unanswered. In response to the appeals to the supportive experimental research, Merikangas and her team said such studies "have no bearing on the results of our meta-analysis" (Merikangas, Lehner, and Risch 2009, 1862). Munafò was willing to grant that such studies "are useful and interesting" (Munafò et al. 2010, e22). But, like Merikangas, he replied that they cannot say whether or not the statistical interaction effect actually exists in natural human populations; only large studies or meta-analyses can weigh in on that. When it came to the studies of alternative environmental stressors (like heart attacks and hurricanes), the meta-analysis designers defended their restrictive approach; advocates of the gene-environment interaction approach were running much too loose with what counted as a "replication," as the title of Munafò's reply—"Defining Replication"—warned (ibid.). Laramie Duncan and Matthew Keller also defended the restrictive approach taken by Merikangas and Munafò (Duncan and Keller 2011). They distinguished "direct replications" from "indirect replications," linking what I am calling the restrictive approach to inclusion of only direct replications and what I am calling the permissive approach to inclusion of direct and indirect replications. Sen's meta-analysis, they argued, employed "extremely liberal inclusion criteria" because it included indirect replications, while Merikangas's and Munafò's meta-analyses appropriately included only direct replications (ibid., 1047).

The Explanatory Divide in the Twenty-First Century: "The Origins Can Be Found in the Famous Dispute between Fisher and Hogben"

This review of the dueling meta-analyses brings us back to where we started—Hardy and Low's question: "What should I believe? Which explanation is true?" The question about which explanation is "true" is ultimately an empirical matter, which I will take up in chapter 6. But the question about which explanation should be "believed" is an epistemological matter about which meta-analysis warrants belief. I will attend to this question here. The answer is: *either, depending on which side of the explanatory divide you fall*. Let me explain by, first, recalling what the explanatory divide is and how it played out in the previous chapters, and then, second, showing how it played out in this most recent debate over the serotonin transporter gene, stress, and depression.

In chapter 2, remember, R. A. Fisher and Lancelot Hogben faced off over three questions in the midst of the early-twentieth-century eugenics controversy: the conceptual question (what is interaction?), the investigative question (why and how should interaction be investigated?), and the evidential question (what is the empirical evidence for interaction?). For Fisher, interaction was a statistical measure of the breakdown in additivity between the main effects of nature and nurture. Conceptualized as such, interaction was to be investigated to the extent that it presented a potential problem for statistical attempts to partition the causes of variation in a population. But, for Fisher, he took the problem to be only a potential once, since cases of interaction were rare in nature and, if they did arise, could be eliminated with a transformation of scale. For Hogben, in contrast, interaction was variation that resulted from differences in unique, developmental combinations of nature and nurture. Conceptualized as such, interaction was to be sought out because of the important information it provided about the causal mechanisms of the developmental process. For Hogben, since interaction was developmental in nature, it would be the rule, not the exception, in nature.

In chapter 3, I traced Fisher's answers to the conceptual, investigative, and evidential questions through mid-twentieth-century agricultural geneticist Jay Lush to Arthur Jensen, who kicked off controversy with his appeal to genetic differences as the source of differences in IQ between African American and Caucasian populations. Jensen understood interaction to be a purely statistical concept that created a potential obstacle to his statistics, but it was only that. In contrast, I traced Hogben's answers to those questions through mid-twentieth-century developmental geneticist Conrad Hal Waddington to Richard Lewontin, who pointed to interaction as fundamentally undermining Jensen's thesis; interaction, developmentally understood, was common in nature and should be sought out for the information it supplied about the causal mechanisms of the developmental process.

The most recent debate over the serotonin transporter gene, stress, and depression specifically, and the more general debate over whether to opt for a genome-wide association study approach or a gene-environment interaction approach carries forward the elements of the explanatory divide that have separated these two ways of thinking about interaction for nearly 100 years now. The genome-wide association study approach, favored by the likes of Merikangas and Munafò, is the modern-day exemplar of what I have been calling the variation-partitioning approach to studying the relationship between nature and nurture. Administers of the genome-wide approach, like designers of heritability studies before them (discussed in the last chapter), focus on measuring genetic causes of variation, determining how much of that genetic variation accounts for the total phenotypic variation under investigation. This focus on answering how-much questions about the causes of variation responsible for variation in a population and utilizing the statistical genome-wide association study approach to answering those questions places these scientists in a tradition that runs back through Jensen, Lush, and Fisher (refer back to table 2.1 in chapter 2 for a tabular presentation of the components of the variation-partitioning approach).

In contrast, Moffitt and Caspi's repeated appeals to the experimental research that manipulated the mechanisms of stress response situates them clearly in an approach to studying nature and nurture that runs back through Lewontin, Waddington, and Hogben: the mechanism-elucidation approach. Their research on interaction paves the way for follow-up research designed "to unravel the mechanisms" of stress vulnerability—to determine how genetic differences and environmental differences relate during development to lead to differences in depression (Kaufman et al. 2010, e19). The experiments on human and non-human animals, they counsel, "elucidate biological mechanisms behind the [gene-environment interaction] hypothesis" (Caspi et al. 2010, 510). The positive results of an interaction effect across a range of environmental stressors point to a "unifying mechanism" that pulls together the diverse outcomes (ibid., 515). Moffitt and Caspi's studies of gene-environment interaction are the modern-day exemplar of the mechanism-elucidation approach to understanding interaction (refer back to table 2.1 in chapter 2 for a tabular presentation of the components of this approach).

Interaction: In Obstruction to Variation-Partitioning or in Service of Mechanism-Elucidation?

Rutter, who brought Moffitt and Caspi to the SGDP and subsequently co-authored with them, has explicitly situated Moffitt and Caspi's research as well as his own in a tradition that traces its route back to Hogben (Moffitt, Caspi, Rutter, and Silva 2001; Moffitt, Caspi, and Rutter 2005; Rutter, Moffitt, and Caspi 2006). "For both [Hogben and Waddington]," Rutter emphasized positively, "G×E was not a nuisance term that needed to be removed by scaling modifications but, equally, it was not an answer in its own right. Rather, as others have argued subsequently (Rutter and Pickles 1991), the finding of G×E points to a phenomenon that requires a biological understanding"[11] (Rutter 2008, 969). Moreover, Rutter saw Moffitt and Caspi's interaction result for the serotonin transporter gene, stress, and depression in precisely this light of mechanism-elucidation:

The findings fall short of showing exactly what neural mechanism is implicated, but they show where the "action" is taking place and at a different form of basic science, studying mechanisms at the cellular level, is required in order to proceed down the path of tying down the precise neural mechanism. (ibid., 973)

Rutter has also contrasted the Moffitt-Caspi approach to studying gene-environment interaction, which traces its roots to Hogben, with critics of this approach, who trace their roots back to Fisher.

The most obvious and straightforward reason [critics have expressed extreme skepticism about the existence of interaction] is that they have focused exclusively on G×E as a statistical concept concerned only with a multiplicative, synergistic interaction using a logarithmic scale. The origins can be found in the famous dispute between Fisher and Hogben. Fisher, the originator of much of the modern statistics, tended to regard G×E as a 'nuisance' term to be eliminated by means of appropriate statistical manipulations, in order to get on with the serious business of partitioning the population variance into genetic and environmental components. Hogben, a biologist as well as a statistical mathematician, argued that the focus had to be on biological interactions if there were to be health benefits deriving from an understanding of the biological pathways involved. (Rutter 2010, 2; see also Rutter, Thapar, and Pickles 2009)

What Rutter points to here is the fact that Moffitt and Caspi's critics, and particularly those critics who helped design the Merikangas and Munafò meta-analyses, have deployed arguments against gene-environment interaction that bear a striking resemblance to the arguments deployed earlier by Fisher against Hogben and Jensen against Lewontin. Take Lindon Eaves, a past president of the Behavior Genetics Association and one of the conceptualizers and designers of the Merikangas meta-analysis. In 2006, the same year as the NIMH workshop that led to the meta-analysis, Eaves questioned whether Moffitt and Caspi's results were fact or artifact (Eaves 2006). Eaves was particularly critical of the possibility that gene-environment interaction results arise from a pathology of scale, rather than from the actual neurobiology of genes and the environment. "Although interactions are widespread in experimental organisms," he granted, "their contribution is typically smaller than those of main effects. More importantly, even when quantitative traits are considered, the effects of G×E can be simulated by problems of measurement and, in some circumstances, can be generated or removed at will by a simple transformation of scale" (Eaves 2006, 1). Eaves's appeal to the small size of interaction effects, and a transformation of scale to make interaction go away, was precisely Fisher's reply to Hogben when it came to cases of interaction. Eaves also, already in 2006, was critical of the "new paradigm for research in psychiatric genetics" presented by Moffitt and Caspi's gene-environment interaction approach, a sentiment that was to be repeated with Merikangas and the other co-authors three years later when they questioned the "substantial sources" devoted to this new paradigm (ibid., 6).

As another example, take Glyn Lewis, who co-authored the Munafò meta-analysis and was responsible (alongside Munafò) for deciding which studies to include in that meta-analysis. One year later, Lewis co-authored an article entitled "Misconceptions about Gene-Environment Interactions in Psychiatry," and there he warned that "misunderstandings surrounding the study of gene-environment interactions are very common" (Zammit, Owen, and Lewis 2010, 65). More specifically, he admitted that many different genetic and environmental factors contribute to the development of diseases.

> This is sometimes referred to as gene-environment interaction ... and few scientists would dispute that this is a real phenomenon. However, this is *not* the same definition of gene-environment interaction as that reported by studies following analysis of data, which always refers to a measure of statistical interaction. (authors' emphasis, ibid.)

Lewis's distinction between the correct, statistical concept of interaction and a muddle-headed, developmental concept was precisely Jensen's reply to Lewontin when it came to cases of interaction. And Lewis in 2010, in a phrase that should sound familiar now, raised "serious concerns about the value of time and resources spent in these [interaction] endeavours" (ibid.). Rutter is thus right. The twenty-first-century criticisms of Moffitt and Caspi's gene-environment interaction approach to studying nature and nurture bear a conspicuous resemblance to the criticisms of Lewontin and Hogben deployed in the last century.

Now, to be clear, I am not suggesting that the negative meta-analysis designers generally, nor Eaves and Lewis specifically, were simply wheeling out kneejerk reactions to the latest instantiation of arguments for the importance of interaction. I see no evidence of these critics treating gene-environment interaction as a "nuisance." And, indeed, the very fact that these individuals considered, designed, implemented, and then published meta-analyses of a purported case of gene-environment interaction shows how seriously they took the matter. The critics, in short, have come a long way since Jensen during the IQ controversy. But I am saying that the appeals to the small size of interaction effects and the appeals to a transformation of scale fit into a broader tradition, initiated by Fisher, wherein interaction was a purely statistical concept that bore the burden of proving itself as a reality. And I am saying that the charges of misconceptions about interaction are part of that broader tradition. I am also saying that the restrictive approach to study inclusion when it came to designing the meta-analyses and the dismissal of the experimental research supportive of the original result fit perfectly into this broader tradition. And, finally, I am saying that the dismissive criticisms of the time and resources spent on the "new approach," the "new paradigm," and "these endeavours" fit neatly into this broader tradition that questioned research on and appeals to interaction in nature in favor of an approach that focused instead on purely genetic effects (such as genome-wide association studies).

This is why the answer to Hardy and Low's question ("What should I believe?") is either, depending on which side of the explanatory divide you fall. If, falling on Hogben's side, you think interaction should be sought out for the important information it supplies about the causal mechanisms of the developmental process, then a focus on experimental investigations that probe the causal mechanisms and a permissive approach to study inclusion make perfect sense. Explanation, on this side, is a matter of mechanism-elucidation, wherein the challenge is to determine how genetic differences and environmental differences play out in the developmental process. On this mechanism-elucidation approach, the experimental investigations manipulate the causal variables in the mechanisms in order to determine how they make their differences, and the investigations of novel environmental stressors (included with a permissive approach to meta-analysis design) test how generalizable that proposed mechanism is out there in the world. Moffitt, Caspi, Rutter, and Sen's team clearly fall on this side of the divide.

If, falling on Fisher's side, you think interaction continues to bear the burden of proving itself, then a dismissal of the experimental investigations as irrelevant and a restrictive approach to study inclusion makes perfect sense. Explanation, on this side, is a matter of partitioning variation, wherein the challenge is to determine how much of the assorted causes of variation account for the total variation in a population. On this variation-partitioning approach, embodied at present by genome-wide association studies, the experimental investigations are irrelevant to the how-much question, and the investigations of novel environmental stressors (excluded with a restrictive approach to meta-analysis design) jump the gun by testing the generalizability of an interaction effect before a genome-wide association study has verified the validity of a genetic association in the first place. Eaves, Lewis, and Merikangas's and Munafò's teams clearly fall on this side of the divide. The side of the explanatory divide on which you fall, in short, provides differing warrants for favoring the results of either of the dueling meta-analyses.[12]

This answer, I suspect, will be unsatisfactory for some readers. It will be unsatisfactory if you wanted a conclusive answer to whether or not the serotonin transporter gene and stress really do constitute a case of gene-environment interaction for depression. More generally, it will be unsatisfactory if you wanted a conclusive answer to who ultimately wins the interaction debates: was Hogben or Fisher right about how interaction should be conceptualized? Or Lewontin or Jensen about why and how interaction should be investigated? Were Moffitt and Caspi, or their critics, right about whether or not the serotonin transporter gene and stress provide empirical evidence for interaction?

I have intentionally held off on weighing in on those specific questions in the chapters of part I because, as I said at the outset, the purpose of these historical chapters was to first formulate the precise nature of the explanatory divide and trace the depth of its historical entrenchment, which runs all the way into the present. I wanted to formulate and trace the explanatory divide first, because the thesis of part II is that the explanatory divide can be bridged. And because it can be bridged, it's not simply a matter of picking the winner at each instantiation of the interaction debates. Bridging the explanatory divide and showing how and why each has something to contribute to conceptualizing, investigating, and weighing evidence for interaction is the task of the next two chapters.

II Bridging the Explanatory Divide

Throughout the history of the nature/nurture debate generally and the interaction debates specifically, there have been variation-partitioners and there have been mechanism-elucidators. Variation-partitioners and mechanism-elucidators share a common interest in studying the relationship between nature and nurture, but they go about undertaking that study via two very different explanatory frameworks. Members of the variation-partitioning approach seek to explain variation in a population; they ask *how-much* questions about how much variation is caused by the different causes of variation. Statistical methodologies, such as the analysis of variance, supply the answers about the causes of variation.

Members of the mechanism-elucidation approach seek to explain the developmental process. Mechanism-elucidators ask *how* questions about how differences in nature and how differences in nurture contribute to differences in the causal mechanisms of the developmental process. Experimental methodologies, such as interventionist manipulations, supply the answers about the causal mechanisms.

What is the relationship between the variation-partitioning approach and the mechanism-elucidation approach to studying nature and nurture? There are several possible answers to this question. One possible answer is that one side is right and the other side is wrong. On this account, there is a correct and an incorrect way to study nature and nurture, and the task for the historian or philosopher of science is to wade into the debate in order to commend members of the correct approach and condemn members of the incorrect approach. Another possible answer is that both sides are correct but in a non-overlapping way. On this account, the different approaches to studying the relationship between nature and nurture are so distinct that there is a genuine incommensurability between their questions and their answers, and so the task for the historian or philosopher of science is to disentangle the approaches in order to show

how the members of the different approaches are simply up to different things.

In the next two chapters I will offer a different diagnosis of the relationship between the variation-partitioners and the mechanism-elucidators. Members of the two approaches do indeed identify different things that need explaining, ask different causal questions, appeal to different things that do the explaining, and utilize different methodologies to provide those explanations. But those differences do not isolate variation-partitioners and mechanism-elucidators in two incommensurable worldviews. There is an explanatory divide, but it can be bridged. So the task for me is to construct this explanatory bridge and show how it links up the apparent disconnect between the two approaches. I will undertake this exercise in two steps, by first developing the general explanatory relationship between partitioning variation and elucidating mechanisms (chapter 5), and then by extending that general relationship to the different ways in which members of the two approaches conceptualize, investigate, and judge evidence of interaction (chapter 6). The product will be an integration—of approaches and of concepts of interaction.

5 Population Thinking about Mechanisms: An Integrative Relationship

What is the relationship between statistical studies that investigate how much different causes of variation are responsible for variation in a population, and interventionist studies that investigate how causal mechanisms are responsible for the development of some trait? I am not the first to ask this question. A number of scientists, historians, and philosophers of science have wrestled with the issue. I will start in the next section by highlighting various contributions that have advanced the analysis. Still, I am the first to make the specific distinction between what I have been calling the variation-partitioning and the mechanism-elucidation approaches to studying nature and nurture. This specific distinction, I claim, opens the door to a novel formulation of the relationship therein.

This is not an indictment of those earlier contributions, for the novel formulation that I will offer actually draws upon very recent developments in the philosophy of science. I have access to philosophical tools, that is, that many of the previous writers did not have. Two philosophical tools, to be exact—one tool for each approach. On the side of the mechanism-elucidation approach, the previous historical chapters conveyed the fact that members of that approach continuously sought to investigate the causal mechanisms of the developmental process. This is the first philosophical tool: in the last fifteen years a *philosophy of mechanisms* has emerged in the philosophy of science, which I will claim provides the resources to assess precisely what the members of this approach were seeking. On the side of the variation-partitioning approach, the previous historical chapters conveyed the fact that members of that approach continuously sought to investigate the causes of variation in a population. The second philosophical tool: philosopher of science C. Kenneth Waters (2007) has introduced to the philosophical literature on causation his concept of an *actual difference maker,* which I will claim provides the resources to assess precisely what the members of this approach were seeking.

The challenge of bridging the explanatory divide between the variation-partitioning and the mechanism-elucidation approaches is the challenge of uniting the contributions of the philosophy of mechanisms with Waters's actual difference maker. The structure of the chapter will look like this: after applying the philosophy of mechanisms to the mechanism-elucidation approach ("thinking about mechanisms") and applying the actual difference maker concept to the variation-partitioning approach ("population thinking"), I will formulate the relationship between the two, generating what I call "population thinking about mechanisms." Population thinking about mechanisms is the bridge that spans the explanatory divide.

Exclusivity, Inclusivity, and the Explanatory Wall

Controversial episodes from the history of science, such as the eugenics controversy and the IQ controversy, have left a lasting impression on scientists as well as historians and philosophers of science. Members of these fields have been drawn to these episodes in an attempt to analyze the underlying metaphysical and epistemological nature of the controversies. These attempts have ranged from the exclusive (wherein the aim was to defend one approach and criticize the other) to the inclusive (wherein the aim was to formulate a relationship between the two approaches). I will discuss several such attempts below, starting with the more exclusive attempts and moving to the more inclusive attempts.

Exclusive Attempts

A form of exclusion between the variation-partitioning and the mechanism-elucidation approaches can be historically traced back to the IQ controversy discussed in chapter 3. As told in the section "Explaining the Controversy, Take 1: Ignorance and Bias," a number of historians and philosophers of science followed Richard Lewontin's lead in surmising that the search for causes of variation responsible for variation in a population was in no way related to the search for causal mechanisms responsible for development. Lewontin drew particular attention to the assumed additivity in the analysis of variance, which the interaction of nature and nurture seemingly undermined. Thus, according to this view, the underlying assumption of the variation-partitioning approach simply did not reflect the interactive nature of biological reality.

This critical view has been carried forward by scientists as well as historians and philosophers of science. Developmental psychobiologist Gilbert Gottlieb, for example, has done more than any other individual to reaffirm Lewontin's points from the 1970s in response to the rise and institutionalization of human behavioral genetics (Gottlieb 1992, 1995, 2003; see Lickliter and Logan 2007 as well as the articles associated with that special issue of *Developmental Psychobiology* for reflections on Gottlieb's influence). To recall from chapter 3, behavioral genetics first took shape as a discipline in the 1960s and became institutionalized with its own journals and societies in the 1970s. It is the modern-day standard bearer of the variation-partitioning approach, with its employment of the analysis of variance utilizing data from twin, adoption, and family studies and its more recent turn to genome-wide association studies. Behavioral genetics, Gottlieb argued, is decidedly not developmental: "The population view of behavioral genetics ... is based on the erroneous assumption that the quantitative analysis of the genetic and environmental contributions to individual differences sheds light on the developmental process of individuals" (Gottlieb 2003, 338). Gottlieb's arguments have been reaffirmed by the next generation of developmental psychologists and biologists, such as David Moore (Moore 2002, 2006, 2008a). Moore edited a special issue of *New Ideas in Psychology* dedicated to the memory of Gottlieb, and there he warned, "even though such techniques measure variability across populations, *they provide no understanding at all of the mechanical causes of the individual development that* must *be responsible for that variability*." (author's emphasis, Moore 2008a, 382)

Turning from scientists to historians and philosophers of science, I mentioned a number of scholars in chapter 3 who carried forward Lewontin's criticisms of the variation-partitioning approach. Rather than repeating their names and claims here, I will instead focus on a more recent contribution from Evelyn Fox Keller (2010). Fox Keller's contribution is particularly relevant here for two reasons. First, she and I share the aim of moving past the sociopolitical accusations in order to identify the deeper philosophical issues at stake. "But I am convinced that social and political agendas are not all that is at work," she counsels. "At least equally important is the extent to which we are plagued by a linguistic practice that actively elides the distinction between the causal dynamics of individual development and statistical patterns of correlation, and that, by doing so, structures both our basic intuitions and our reasoning" (ibid., 63). Fox Keller's book is devoted to dissecting those linguistic practices, and she does so via a number of distinctions: individuals versus populations, traits versus trait differences, and how versus how-much questions. So whereas I am focusing on the explanatory divide between the variation-partitioning and mechanism-elucidation approaches, Fox Keller draws attention to the linguistic divide therein.

The second reason Fox Keller's contribution is particularly relevant to my analysis stems from the fact that her book was written and published after Waters introduced his actual difference maker concept, which I will rely on later in this chapter. Thus, Fox Keller had the opportunity to consider Waters's contribution (whereas the other historians and philosophers of science I mentioned in chapter 3 did not). I will detail his contribution below, but for the present purpose it suffices to say that Fox Keller worries that Waters's actual difference maker concept only perpetuates the linguistic confusion that she identifies. For instance, with regard to the distinction between how questions about trait development and how-much questions about trait differences, Fox Keller warns:

> But the problem with which I am concerned arises when we attempt to infer from the answers to that question (how much) an answer to the earlier question of how, because in general, this is not possible. With Waters's interpretation of causal statements of the form 'x causes y' as meaning 'a difference in x causes a difference in y', it becomes all but impossible to mark, or even to recognize, the distinction between the two questions, and correspondingly easy to think that by answering the second, one has also answered the first." (ibid., 39)

Inclusive Attempts

Gottlieb, Moore, and Fox Keller's criticisms of the variation-partitioning approach can be contrasted with the more inclusive attempts of self-described "pluralists." For inclusive pluralists, the aim is not to criticize the variation-partitioning approach but rather to separate the scope of the approaches, thereby limiting the appearance of competition. A recent example of this pluralism can be found in the work of philosopher of science Helen Longino. Longino has spent over a decade utilizing the case of research on aggression to consider the relationship between the various scientific approaches that study human behavior (Longino 2001). In her *Studying Human Behavior: How Scientists Investigate Aggression and Sexuality* (2013), Longino partitions the scope of five of these approaches, noting their respective objects of measurement, methodologies, and domain of causal inquiry. First, quantitative behavioral genetics studies human behavior with data from twin and adoption studies using the analysis of variance and heritability measures; it identifies the causes of variation responsible for variation in populations. Second, social environment approaches study human behavior by focusing on the determinants of behavior in the social environment; in contrast to quantitative behavioral genetics, social environment approaches do not aim to distinguish genetic contributors from environmental contributors, but rather aim to discriminate the contributions made from among the environmental determinants. Molecular behavioral genetics comprises the third approach to studying human behavior. Molecular behavioral genetics utilizes gene-finding methods (such as genome-wide association studies) as well as molecular methods that link particular segments of DNA with functional protein products. Fourth, neurobiological approaches identify the neural structures and processes involved in human behavior; these approaches rely on methods ranging from neural imaging to pharmacological interventions. And finally, integrative approaches combine results from several of the aforementioned approaches. For instance, Longino characterizes the research by Terrie Moffitt and Avshalom Caspi (introduced in chapter 4) as an integrative gene×environment×neurosystem approach, which builds off the results of molecular behavioral genetics, social environment approaches, and neurobiological approaches; despite being integrative, however, Longino notes that this approach is confined to the study of clearly defined or delineated psychiatric disorders.

Unfortunately, Longino points out, these different approaches to studying human behavior can become competitive—for sources of funding, for attention from the media, for fulfilling the promise to "win the race" of understanding human behavior. But this competition belies the fact that the approaches actually overlap surprisingly little. For example, although all of the approaches attempt to answer the question, "what causes human behavior?," in actuality "behavior" means different things in those different approaches: a tendency in a population, an episode in the history of an individual, a pattern of behavior, the disposition to respond to situations in one way or another, or a pattern of interactions (ibid., chapter 7). Likewise, although all of the approaches attempt to answer the question, "what causes human behavior?," in actuality the different approaches focus on very different portions of what Longino calls the "causal space" of human behavior. Figure 5.1 reproduces Longino's potential causal space. Quantitative behavioral genetics, she explains, confines itself to genetic causes (the two boxes on the far left of the figure) and a subset of environmental causes (the single box on the far right of the figure). The social environment approaches, in contrast, confine themselves to the range of environmental causes (the three boxes on the far right of the figure).

Genotype 1	Genotype 2	Intrauterine environment	Physiology	Nonshared environment	Shared (intrafamily) environment	Socioeconomic status
(allele pairs)	(whole genome)		(hormone secretory patterns; neuro-transmitter metabolism) Anatomy (brain structure)	(birth order; differential parental attention; peers)	(parental attitudes regarding descipline; communication styles; abusive/ nonabusive)	(parental income;level of education; race/ethnicity)

Figure 5.1
The potential causal space of human behavior (reproduced from Longino 2013, figure 1a).

Longino's analysis of the various scientific approaches to studying human behavior is inclusive in nature. However, that inclusivity should not be confused with reductivity or integrativity. Longino describes her pluralism as "non-eliminative." Her partitioning of the various approaches is not a means to a reductive or integrative end, wherein the quantitative is reduced to the molecular or the neurobiological is integrated with the social environmental. She explains, "even though there is, from one point of view, a common phenomenon to be understood—behavior, or rather, specific behaviors or behavioral patterns such as aggressivity or sexual orientation—each approach brings with it a prior and distinctive representation of the domain of investigation" (Longino 2013, 132).

Psychiatric geneticist Kenneth Kendler shares Longino's pluralistic vision; however, in contrast to Longino, Kendler does seek integration (Kendler 2005). Kendler's focus is narrower than Longino's; Longino analyzes the broad spectrum of scientific approaches that investigate human behavior, whereas Kendler just focuses on psychiatric genetics, a subdiscipline of behavioral genetics that investigates psychiatric diseases and disorders. Kendler identifies four "paradigms" within this field: basic genetic epidemiology, advanced genetic epidemiology, gene-finding methods, and molecular genetics. Basic genetic epidemiology quantifies the degree to which individual differences in risk to illness result from familial/genetic factors; it draws on twin, adoption, and family studies to identify genetic risk factors. Advanced genetic epidemiology, which builds off the results of basic genetic epidemiology, explores the nature and mode of action of the genetic risk factors by, for example, determining whether males and females are affected differently or determining whether the genetic risk factor affects exposure to environmental factors. The third paradigm—gene-finding methods—seeks to determine the location on the genome of genes that have variants that differentially impact the liability to psychiatric disorders (i.e., susceptibility genes). Finally, molecular genetics, which builds off the results of gene-finding methods, attempts to elucidate the biological mechanisms that run from susceptibility genes to disease or disorder.[1]

Kendler's first and second paradigms (basic and advanced genetic epidemiology) coincide with Longino's quantitative behavioral genetics approach, while his third and fourth paradigms (gene-finding methods and molecular genetics) coincide with Longino's molecular behavioral genetics approach. Too often, Kendler admits alongside Longino, the molecular "gene jocks" (from paradigms 3 and 4) and the epidemiologists who treat genes as "just statistics" (from paradigms 1 and 2) compete. In contrast to Longino, however, Kendler does not take his pluralistic partitioning of paradigms to be ineliminable: "the field of psychiatric genetics would do best to integrate these four paradigms, stressing their relative strengths and limitations. Thus integration can be best done within an overall framework of explanatory pluralism that values a range of reductive explanations across varying levels of biological and psychological complexity" (Kendler 2005, 3). The question, then, is: what is the relationship between paradigms 1 and 2 and paradigms 3 and 4? Kendler answers that the genetic risk factors identified by paradigms 1 and 2 are theoretically simply the statistical signals of the susceptibility genes studied by paradigms 3 and 4. But at a practical level, the answer is "more murky" (ibid., 8). The problem is (a) the incredible complexity of the relationship between DNA and disease or disorder, which may be genuinely intractable, and (b) the fact that a genetic risk factor identified by paradigm 1 may be distributed across many susceptibility genes with small individual effects, making the transition to paradigm 3 daunting at best.

Longino and Kendler share my goal of carving up disciplinary domains with an eye toward assessing the relationships between those domains. We came to that carving exercise, however, by way of different trajectories, so it is not surprising that the products of our carving do not overlap entirely. Longino is interested in research on human behavior generally, and so she casts a wide net, carving up scientific approaches that range from neuroscience to the social sciences. Kendler, in contrast, confines his interest to his own discipline—psychiatric genetics, and so he casts a comparatively small net, carving up the paradigms of that discipline. My focus is on the debates over the interaction of nature and nurture, and so I cast a net that distinguishes the variation-partitioning approach from the mechanism-elucidation approach, because it is the members of these approaches who consistently faced off in the interaction debates. My focus puts me (in terms of net size) somewhere between Longino's effort and Kendler's effort—more extensive than Kendler's exclusive focus on psychiatric genetics but no so encompassing as Longino's consideration of human behavior research broadly construed. This issue of scope will be important when I return to these authors later in the chapter.

The Explanatory Wall

The goal of this chapter is to bridge the explanatory divide between the variation-partitioning and the mechanism-elucidation approaches. The authors above, however, present a diversity of challenges to such a goal. From the exclusive criticisms of the variation-partitioning approach by Gottlieb, Moore, and Fox Keller, I am challenged to answer the following questions: since the study of populations is not the study of development, how can the study of populations shed any light on the developmental process? Since it is widely known that correlation is not causation, how can the study of causes of variation provide any understanding of causal mechanisms? And since how-much questions are about trait differences, while how questions are about trait development, how can answers to the former say anything at all about answers to the latter?

Longino offers an inclusive, pluralistic vision of the scientific study of human behavior. But her pluralism is non-eliminative in nature, so if I want to bridge the explanatory divide between the variation-partitioning and the mechanism-elucidation approaches, then I am also challenged to answer her question: since the different approaches investigate different portions of the causal space of human behavior, how can those approaches be bridged in any way?

These four authors can be understood as identifying an explanatory wall between the variation-partitioning approach and the mechanism-elucidation approach. Thus, if I want to bridge the explanatory divide between the approaches, then I must scale the explanatory wall that these authors have constructed. Of the authors surveyed, only Kendler shares my integrative vision, but even he warns that it is "murky." I do believe the challenging questions above can be answered in a non-murky way. The answer comes from drawing on two philosophical tools—the philosophy of mechanisms, and the concept of an actual difference maker. The union of the philosophy of mechanisms with actual difference makers results in population thinking about mechanisms. With my formulation of population thinking about mechanisms in place, I will return to the challenges above to show how the explanatory wall can be scaled.

Thinking about Mechanisms

Scientists from (and defenders of) the mechanism-elucidation approach were united in their emphasis on searching for the causal mechanisms of the developmental process. Why? In the last several decades, philosophers have returned to the concept of a mechanism to make sense of why scientists seek mechanisms. (I say "return" to contrast the more recent endeavors from the classic mechanical philosophy of the seventeenth century.) The origins of this effort can be found in the work of William Wimsatt (1976), Wesley Salmon (1984), Kenneth Schaffner (1993), William Bechtel and Robert Richardson (1993), and Stuart Glennan (1996). These philosophers all began considering mechanisms as the source of scientific explanations out of a dissatisfaction with the then-dominant philosophical idea that scientific explanations were based on derivation from laws of nature (Hempel and Oppenheim 1948). This philosophy of mechanisms hit its stride at the turn of the twenty-first century, when a series of philosophers moved beyond specific sciences and assessed the concept of a mechanism generally (Machamer, Darden, and Craver 2000; Glennan 2002). Peter Machamer, Lindley Darden, and Carl Craver's "Thinking about Mechanisms" (2000) has been especially influential. I have titled this section of the chapter "thinking about mechanisms" in part to reference that work and also because "thinking" acts as a terminological bridge between this section and the next, which is itself symbolic of the explanatory bridge I aim to build in this chapter.

There are differences between the various accounts of a mechanism. But, for our purposes, we can focus on what they hold in common—the idea that a scientist explains a phenomenon by identifying and manipulating the variables in the mechanisms responsible for that phenomenon, thereby determining how those variables are situated in and make a difference in the mechanism; the explanation ultimately amounts to the elucidation of how those variables act and interact in the mechanism to produce the phenomenon under investigation. The philosophy of mechanisms is meant to capture how scientists answer questions such as the following: how does the brain relay neural messages across a synapse (Machamer, Darden, and Craver 2000)? How do plants convert solar energy into chemical energy (Tabery 2004b)? How does the cell produce proteins (Darden 2006)? And how does the cell generate energy (Bechtel 2006)? Such questions are answered by elucidating the mechanisms responsible for synapse transmission, photosynthesis, protein synthesis, and adenosine triphosphate synthesis.

A detailed example will help to show how this actually works. Craver, in his *Explaining the Brain* (2007), shows how neuroscientists explained a phenomenon like spatial memory by elucidating the multilevel mechanisms responsible for it (see figure 5.2). Spatial memory is the portion of memory devoted to one's environment and one's orientation in that environment; neuroscientists elucidated the multilevel mechanisms responsible for spatial memory largely via research on rats (hence the rat at the top of Craver's figure), but spatial memory and the mechanisms of spatial memory are highly conserved, suggesting that the explanation of spatial memory in the rat can be extrapolated to other organisms such as humans (see Ankeny 2000, Schaffner 2001, Weber 2005, Steel 2008, Nelson 2013 and Piotrowska 2012 on philosophical issues surrounding extrapolation). At the organismic level of spatial memory, the rat is placed in a variety of mazes, and researchers test spatial memory by timing how long it takes for the rat to complete the mazes. At the brain-system level, spatial memory is believed to be generated by the hippocampus, which constructs spatial maps. At the lower cellular level, spatial map generation is understood to be achieved via the long term potentiation (LTP) of neurons in the hippocampus. Finally, at the molecular level, LTP is made possible by molecular mechanisms, such as *N*-methyl-D-aspartate (NMDA) receptors, that control neurotransmission.

For neuroscientists to elucidate the multilevel mechanisms responsible for spatial memory, they had to account for both (a) causal relations and (b) constitutive relations. Craver draws on the work of James Woodward (2003) to make sense of the former. Woodward's manipulationist account of causal explanation tightly links explanation and manipulation. Scientists causally explain when they know how to manipulate. Manipulations are understood counterfactually; if some particular variable is a cause of some outcome, then manipulating the value of that variable would be a way of manipulating the outcome. These counterfactual experiments formulate and then answer what-if-things-had-been-different questions; and, in so doing, they establish a pattern of counterfactual dependence.[2] In sum, to say that there is a causal relationship between two variables (X and Y), according to Woodward, is to say that one can change the value (or probability distribution) of one variable (Y) by manipulating the other variable (X). Craver points out that this is precisely what neuroscientists did to

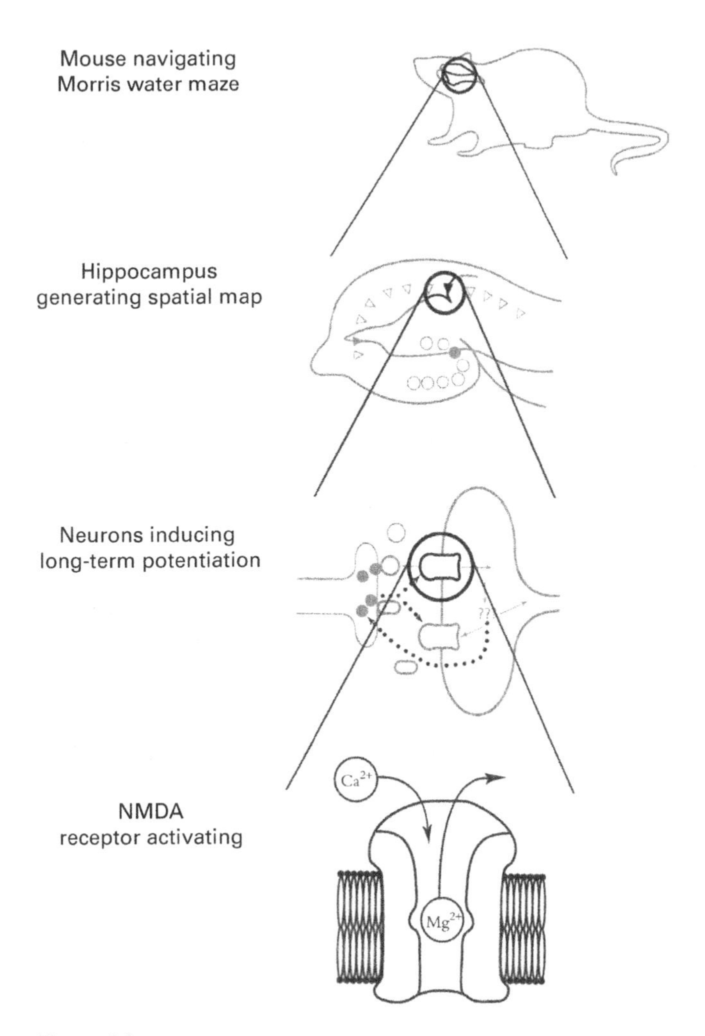

Figure 5.2
The levels of spatial memory (reproduced from Craver 2007, figure 5.1, by permission of Oxford University Press).

identify causal relations involved in the mechanisms of spatial memory. For example, experiments that inhibited or strengthened LTP led to poorer or better spatial memory; in this way, the causal relation between LTP and spatial memory was established. Likewise, experiments that altered the NMDA receptor (by, for example, changing its conformation) led to compromised LTP, spatial map formation, and maze performance; in this way, the causal relation between the NMDA receptor and the higher levels was established.

Woodward's manipulationist account of causal explanation captures neuroscientists' establishment of causal relations quite nicely, but that still leaves out constitutive relations between levels. Figure 5.2 depicts four levels of spatial memory in the rat, but what are these levels? Philosophers have applied level-talk to a variety of phenomena—levels of science, levels of size, levels of aggregativity. The levels of spatial memory, Craver argues, are levels of mechanism. Levels of mechanism are compositional in nature, wherein lower-level mechanisms constitute components of higher-level mechanisms. Craver's depiction of the levels of spatial memory in the rat clearly shows this constitutive relationship between levels: the NMDA receptor is a molecular mechanism responsible for the flow of calcium ions into the post-synaptic neuron; that molecular mechanism is a component of the cellular mechanism of LTP responsible for strengthening the efficiency of neurotransmission; that cellular mechanism is a component of the hippocampal mechanism responsible for forming spatial maps; and that hippocampal mechanism is a component of the rat involved in navigating the maze.

Typological Thinking about Mechanisms

Notice that the neuroscientists treated "the rat" in the previous spatial memory example quite generally. The spatial memory example was not about a specific rat or even about a population of rats. It was about "the rat." The experiments performed to elucidate the various mechanisms at the various levels likely involved thousands of rats, but the particularities of each individual rat were not of interest. Rather, the results from the various experiments on the various rats were compiled to explain spatial memory in what philosophers would call "the Platonic form of the rat."[3]

The eminent evolutionary biologist Ernst Mayr referred to such generalizing by scientists as "typological thinking" (Mayr 1975). But he introduced this characterization of one form of scientific thinking primarily to contrast it with what he called "population thinking." Population thinking, which Mayr traced to Charles Darwin, was quite different from typological thinking in what the scientists focused on and what they ignored. "The ultimate conclusion of the population thinker and of the typologist are precisely the opposite," Mayr explained. "For the typologist, the type (*eidos*) is real and the variation an illusion, for the populationist the type (average) is an abstraction and only the variation is real. No two ways of looking at nature could be more different" (Mayr 1975, 27). Mayr himself recognized the Platonic nature of typological thinking; the typologist's "eidos" is a Platonic concept from Plato's theory of forms.[4]

The thinking about mechanisms undertaken by philosophers of mechanisms thus far has been typological thinking about mechanisms. This is because these philosophers of mechanisms have drawn their case studies from typological scientists who, when studying a phenomenon, tend to ignore variation and focus instead on the generalizations that can be drawn across individuals. Craver's example of spatial memory in "the rat" is one instance, but it is representative of the philosophy of mechanisms broadly, which points to typological research on neurotransmission, photosynthesis, and protein synthesis as case studies depicting how scientists explain. If scientists only engaged in typological thinking, then this singular attention to typological thinking about mechanisms would be all well and good. But, as Mayr stressed, scientists also engage in population thinking.[5]

Population Thinking

Members of the variation-partitioning approach, I claim, are population thinkers. They study the causes of variation responsible for variation in populations. What, then, *is* a "cause of variation"? R. A. Fisher first introduced "cause of variation" in his landmark 1918 paper (discussed in chapter 2), and the causal language has been situated at the heart of the variation-partitioning approach ever since (Fisher 1918). But this historical point only tells us from where the term came; it doesn't tell us what a cause of variation actually is. The task of this section is to answer: what is a cause of variation?

The idea of a causal mechanism is much more intuitive. Suppose, for example, I point to a clock on the wall and ask you, "What is the causal mechanism responsible for the time displayed on that clock face?" You would answer that there are an assortment of cogs, gears, and springs that are tuned just so to move the hour and minute hands around the clock face in order to tell time. But what if I then take you to a clock shop where 20 of the same clock are all telling a slightly different time. I now ask you,

"What is the cause of variation responsible for the variation in times displayed on those clocks?" It is less obvious how to answer this question. But it is precisely this type of question about causes of variation that we must make sense of if we are to make sense of the variation-partitioning approach. A paper by C. Kenneth Waters provides the resource.

Consider two situations: first, Mary strikes a match, which lights it. Second, a segment of DNA is transcribed and translated to produce a protein. In both situations, you might be inclined to say that Mary's striking of the match was "the cause" of its lighting, and that the particular order of nucleotide base pairs in the DNA was "the cause" of the particular protein. But this analysis of the situations ignores the fact that there were many other causes also present and contributing to both situations. In the case of the lit match, there had to be oxygen in the room for the match to catch fire. Oxygen was also a cause of the lighting. In the case of the protein, there had to be a host of other molecules contributing to the process of protein synthesis, such as RNA polymerase and ribosomes. These molecules were also causes of the protein product. So is there any sense in which you can actually say that one cause among many is "the cause"? For hundreds of years, philosophers have answered, "No." We may pick out the striking of the match or the segment of DNA as "the cause," the argument goes, but this picking out is an inherently interest-laden enterprise that reflects our particular interests rather than something ontological about the world.[6]

Waters challenges this received view by drawing on the biological research of early-twentieth-century geneticist Thomas Hunt Morgan. Morgan and his colleagues bred fruit flies to investigate how particular traits, such as eye color, were transmitted from one generation to the next. They did this by constructing experimental situations that ensured the difference in a particular phenotypic trait like eye color (red versus purple) came about due to the difference in a single gene (the purple eye gene). Waters points to a thought experiment in which Morgan asks his readers to imagine a scenario where every gene in the fruit flies' genome contributes to eye color, but only one gene is actually changed to actually change eye color (from red to purple). Can we say that the one gene was "the cause" of the effect (having purple eyes)? "In a strictly causal sense this is true," Morgan admits, "but the effect is produced only in conjunction with all the other genes. In other words, they are all still contributing, as before, to the end-result which is different in so far as one of them is different" (cited in Waters 2007, 559). This quote, Waters emphasizes, shows that Morgan recognized many genes could contribute to a phenotypic trait; and yet, if the change in eye color was brought about by the change in a single gene, then "in a strictly causal sense" that single gene could be identified as the actual cause of the effect. The question for Waters is: how could Morgan simultaneously recognize (a) the contributions of many causes and yet (b) single out an actual cause?

Waters, like Craver, draws on Woodward's manipulability theory of causation and causal explanation as a starting-point to answer his question. Recall that to say there is a causal relationship between two variables (X and Y), according to Woodward, is to say that one can change the value (or probability distribution) of one variable (Y) by manipulating the other variable (X). Woodward's theory, Waters points out, captures quite naturally the causal relationships identified by Morgan. "For at least some organisms, a manipulation of the form of a gene they possessed as a zygote, given other appropriate conditions (such as genetic background and environmental conditions), would change the phenotypes of those individuals" (ibid., 564). Woodward's theory, however, only goes part of the way toward answering the question about simultaneously recognizing (a) the contributions of many causes while (b) still singling out the actual cause of an effect. Go back to Morgan's thought experiment to see why: if every gene in the fruit flies' genome contributes to eye color, then every gene could potentially be manipulated to change eye color. All Woodward supplies us with is an understanding of the first half of Morgan's statement—that every gene makes a causal contribution to eye color. Every gene, that is, is a *potential* difference maker. But only one gene actually made the difference. "In a strictly causal sense," Waters stresses, only one gene was the *actual* difference maker; that is, "in order for there to be a difference maker there must be a difference. And in order for there to be an *actual* difference maker, there must be an *actual* difference" (ibid., 566).

If the actual effect in question is caused by a single actual difference, then we may speak of *the* actual difference maker. Waters defines "the actual difference maker" as follows (ibid., 567):

1. X causes Y (in the sense of Woodward's manipulability theory).
2. The value of Y actually varies among individuals in population p.
3. The relationship expressed by "X causes Y" is invariant with respect to the variables that actually vary in population p (over the spaces of values those variables actually take in p).
4. Actual variation in the value of X fully accounts for the actual variation in Y values in population p (via the relationship X causes Y).

So even if every gene contributes to eye color and every gene is a potential difference maker, in the population only a single gene actually differed (the purple eye gene). The variation in the purple eye gene, that is, fully accounted for the actual variation in the red versus purple eye color in the population. Thus, the purple eye gene was the actual difference maker. This same analysis can be applied to Mary's match and the DNA from earlier. In the case of the lit match, oxygen was a potential difference maker, but it was Mary's striking of the match that actually made the difference, and so Mary's striking of the match was the actual difference maker. Likewise, in the case of protein synthesis, if RNA polymerase and ribosomes do not actually vary across a population but DNA does, then RNA polymerase and ribosomes can both be counted as potential difference makers, but DNA can be counted as the actual difference maker because it is the difference in DNA that actually makes the difference in the protein product.[7]

Waters is quick to point out that in many biological situations (especially those outside the confines of experimentally controllable environments) there is not just one actual difference maker; rather, there are many. In such situations, "an" actual difference maker replaces "the" actual difference maker. Waters defines "an actual difference maker" as follows (ibid., 571):

1. X causes Y (in the sense of Woodward's manipulability theory).
2. The value of Y actually varies among individuals in population p.
3. The relationship expressed by "X causes Y" is invariant over at least parts of the space(s) of values that other variables actually take in population p. (In other words, it is invariant with respect to a portion of the combinations of values the variables actually take in p.)
4. Actual variation in the value of X partially accounts for the actual variation of Y values in population p (via the relationship X causes Y).

The difference between "the actual difference maker" and "an actual difference maker" is that the former fully accounts for the actual variation in effect, whereas the latter only partially accounts for the actual variation in effect.

With Waters's contribution in place, we may now return to the question that opened this section: what is a cause of variation? Answer: an actual difference maker is a cause of variation. To see this, consider the example from chapter two in which Fisher first considered multiple causes of variation. In his study of 12 different potato varieties' responses to six different fertilizers, Fisher investigated how much of the total variation in weight of produce from the population could be attributed to potato variety and fertilizer (Fisher and Mackenzie 1923). In Waters's terminology, Fisher was investigating how much of the actual difference in weight of produce could be attributed to the actual difference in potato variety and the actual difference in fertilizer. The analysis of variance table included in Fisher's paper assigned portions of the total variation to the various sources; fertilizer accounted for most of the variation, and potato variety accounted for some as well. In other words, fertilizer and potato variety were both actual difference makers in the population, having both partially accounted for the variation in weight of produce. (If, on the other hand, Fisher found fertilizer alone fully accounted for the variation in weight of produce, then fertilizer would have been the actual difference maker.)

Much of the variation-partitioning approach, from Fisher to Jensen to modern behavioral genetics, should be understood as the search for actual difference makers. The search for the causes of variation responsible for variation in a population is the search for actual difference makers responsible for actual difference in a population. And the associated how-much question is about how much of the actual difference is attributable to the actual difference makers. Now, the variation-partitioning focus on actual difference makers can (and largely did) proceed in an agnostic fashion regarding the causal mechanisms that linked actual difference makers to actual differences. (Waters notes that Morgan similarly did not need to concern himself with the causal mechanisms of gene action in his search for actual difference makers.) By focusing on the statistical how-much question, members of the variation-partitioning approach largely ignored or "black-boxed" the underlying causal mechanisms. This historical fact about agnosticism concerning causal mechanisms, however, should not be confused with an epistemological necessity. For, as I will argue in the next section, population thinking about actual difference makers can be incorporated into the search for and elucidation of causal mechanisms.

Population Thinking about Mechanisms

I introduced the philosophy of mechanisms in this chapter because that philosophy captures nicely the focus of the mechanism-elucidation approach on the search for the causal mechanisms of the developmental process. And I introduced Waters's actual difference maker concept because that concept captures nicely the focus of the variation-partitioning approach on the search for the causes of variation responsible for variation in a population. Bringing the two approaches together, then, is a matter of bringing the philosophy of mechanisms and Waters's actual difference maker concept together. That union is what I call "population thinking about mechanisms," which I argue is the explanatory bridge between the variation-partitioning and mechanism-elucidation approaches. To show how population thinking about mechanisms works, I will start by way of an example—research on brain-derived neurotrophic factor (or BDNF). BDNF is relevant here, first, because it figures into the molecular mechanisms of spatial memory (Craver's example) and, second, because the gene coding BDNF is an actual difference maker (Waters's concept). With the scientific research on BDNF in place, I will then use that example to show why population thinking about mechanisms is the explanatory bridge between (a) variation in populations and the developmental process, (b) how-much questions and how questions, (c) causes of variation and causal mechanisms, and (d) statistical and interventionist methodologies.[8]

Brain-Derived Neurotrophic Factor and the Molecular Mechanisms Regulating NMDA Activity

Brain-derived neurotrophic factor is one type of neurotrophin, a family of proteins involved in the development and function of neurons. In the 1980s, Yves-Alain Barde at the Max Planck Institute for Psychiatry in Germany led research that first identified and purified BDNF (Barde, Edgar, and Thoenen 1982; Leibrock et al. 1989; Barde 1989). Throughout the 1990s, a series of studies led by Eric Levine of the University of Medicine and Dentistry of New Jersey revealed BDNF to be a crucial entity in the levels of mechanisms responsible for spatial memory. For example, Levine and his colleagues took slices of hippocampal brain tissue and found that adding BDNF rapidly (over minutes) increased synaptic transmission efficacy (Levine et al. 1995; see also Kang and Schuman 1995). This is an example of what Craver calls an interlevel, bottom-up "stimulation experiment" (Craver 2007, 149). Interlevel, bottom-up "interference experiments" also took place (ibid., 147); for instance, BDNF knockout mice were created that lacked BDNF and that, in turn, suffered severely compromised LTP. However, when BDNF was added back to the hippocampal slices from these knockout mice, LTP was completely restored (Patterson et al. 1996). Levine and his colleagues (1998) also localized the role of BDNF as acting on the NMDA receptor (the key entity at the bottom, molecular-mechanism level of Craver's figure 5.2). Specifically, the binding of BDNF to its TrkB receptor (another receptor on the membrane of the postsynaptic neuron) phosphorylated (or chemically added a phosphate group) to the NMDA receptor, which increased the probability of the NMDA receptor being open and, thus, active.

In the 2000s, Toshitaka Nabeshima and his colleagues, of Nagoya University in Japan, linked BDNF up to the spatial memory level and also elucidated the final piece of the molecular mechanism connecting BDNF to the NMDA receptor. To link BDNF up to the spatial memory level, they undertook what Craver calls an "activation experiment"—a top-down experiment wherein the phenomenon to be explained is manipulated to see if the lower-level mechanisms are manipulated in turn. Nabeshima and his colleagues placed rats in maze tests (the spatial memory activation) and then detected changes in BDNF expression in response to this learning exercise (the lower-level detection); they found that BDNF expression did indeed change in response to that spatial learning exercise. Moreover, the researchers also inhibited BDNF expression (an interference experiment) in the rats; doing so impaired both the ability to form new spatial memories and to recall previously learned spatial memories (Mizuno et al. 2000). Three years later, Nabeshima and his colleagues elucidated the mechanism whereby BDNF phosphorylates the NMDA receptor (Mizuno et al. 2003). Fyn, a tyrosine kinase (an enzyme that transfers phosphate groups), acted as the chemical bridge between the BDNF-activated TrkB receptor and the NMDA receptor.

In light of the research described above, scientists have now elucidated a mechanism sketch hypothesizing how BDNF regulates NMDA-receptor activity and how that regulation permeates up through the levels of spatial memory (see figure 5.3). At the molecular level, BDNF binds to the TrkB receptor on the post-synaptic neuron, which in turn phosphorylates the NMDA receptor by way of Fyn. This phosphorylation activates the NMDA receptor, allowing calcium to enter and potentiate the synapse for LTP. The stimulation, interference, and activation experiments by Levine, Nabeshima, and their colleagues showed how the results of that molecular mechanism subsequently affected the cellular, hippocampal, and organismal levels of spatial memory.[9]

The *BDNF* Gene: An *Actual* Difference Maker

Craver's story about the research on spatial memory, supplemented by the additional BDNF work, reveals that scientists elucidated the various mechanisms of spatial memory by way of experimental interventions on various variables (e.g., the NMDA receptor, the hippocampus, BDNF, Fyn). These interventions established the causal relationships between the various variables and the outcome of interest—spatial memory. The establishment of the causal relationships in turn confirmed these various variables to be potential difference makers. The NMDA receptor, for example,

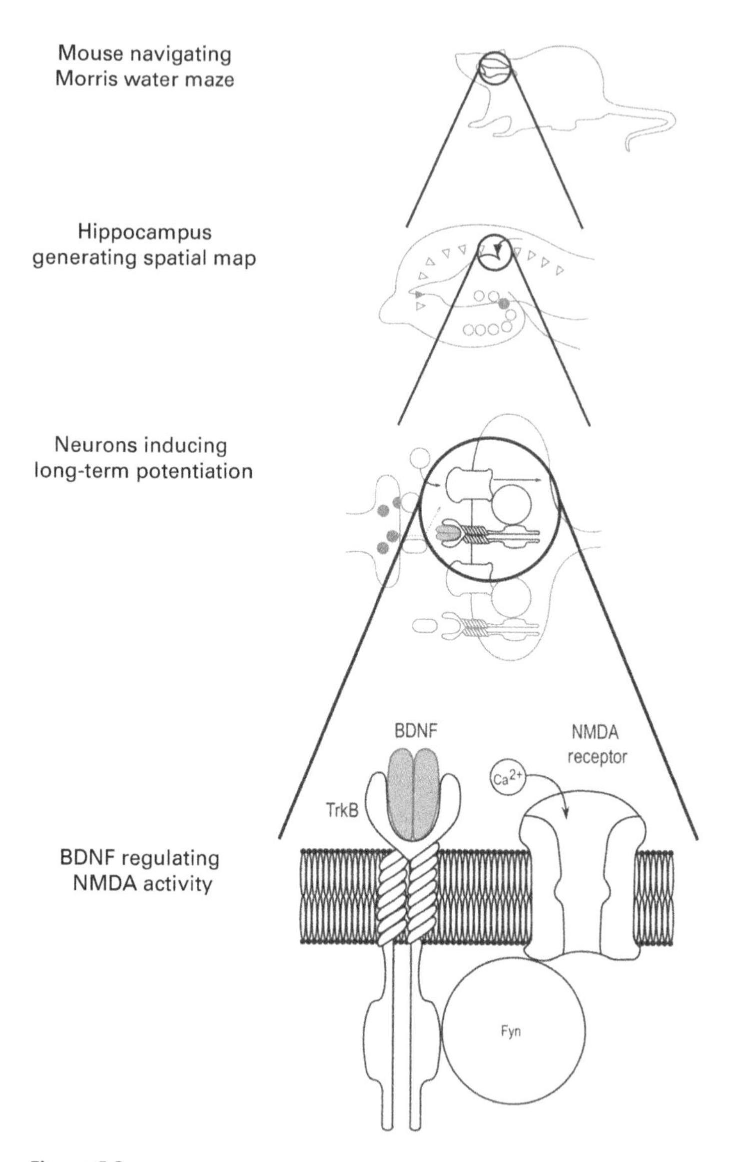

Figure 5.3
BDNF-mediated NMDA activity embedded in the levels of spatial memory (adapted from Craver 2007, figure 5.1, and Blum and Konnerth 2005, figure 4a).

was potentially manipulable by way of experimentally changing its conformation, which in turn made a difference in spatial memory.

In 2003, the same year that Nabeshima identified Fyn as the potential difference maker between BDNF and the NMDA receptor, researchers at the National Institutes of Health, led by Daniel Weinberger, identified an actual difference maker in the molecular mechanism (Egan et al. 2003; Hariri et al. 2003). In previous years, behavioral geneticists had conducted a variety of epidemiological twin, adoption, and family studies to investigate the causes of variation responsible for variation in memory generally and spatial memory specifically (see, for example, Nichols 1978; McGue and Bouchard 1989; Plomin, DeFries, and McClearn 1990). Such studies, however, only pointed to the genome as a whole as a cause of variation. The challenge was transitioning from the entire genome to individual genes linked to specific proteins with known functionality. In the early 1990s, the gene that codes for BDNF (the *BDNF* gene) was identified and located on chromosome 11; this genetic research also confirmed that the *BDNF* gene was conserved among mammals (Jones and Reichardt 1990; Maisonpierre et al. 1991). Behavioral geneticists, not surprisingly, took an interest in the *BDNF* gene and its implications for memory (Dragunow 1996; Morley and Montgomery 2001), ultimately identifying a single-nucleotide polymorphism (or "SNP") at nucleotide 196 in the *BDNF* gene. A SNP is a location in DNA where a single nucleotide differs between individuals. It was the existence of this SNP, along with the role of the BDNF protein in the levels of mechanism responsible for memory, that brought the attention of Weinberger's team to the *BDNF* gene.[10]

The SNP of interest to Weinberger's team was the Val66Met SNP. At nucleotide 196 in the *BDNF* gene, individuals have either a guanine nucleotide or an adenine nucleotide. This 196th nucleotide resides in the 66th codon of the *BDNF* gene; having guanine results in the synthesis of valine (Val) in the amino acid chain, while having adenine instead results in the synthesis of methionine (Met) in the amino acid chain (hence the name "Val66Met"). Individuals receive one copy of the *BDNF* gene from each of their parents, so the possibilities are Val/Val, Val/Met, or Met/Met. In the U.S. population, the Val/Val variant is the most common (68.4%), followed by the Val/Met variant (27.1%), and the relatively rare Met/Met variant (4.5%); however, these frequencies differ in other countries, such as Japan and Italy, where the Val/Val variant frequency is lower (33.8% and 48.7% respectively) and the Met/Met variant frequency is higher (15.9% and 8.1% respectively; Shimizu, Hashimoto, and Iyo 2004).

In the first study from Weinberger's team, the researchers combined rat and human studies to investigate the role of the Val66Met SNP at the molecular, cellular, and hippocampal levels (Egan et al. 2003). At the molecular-cellular level, the researchers employed a bottom-up stimulation experiment by introducing (or transfecting) either the val-BDNF or met-BDNF nucleotide sequence into rat hippocampal neurons and then visually monitoring the intracellular distribution of BDNF by these neurons using green fluorescence protein. The val-BDNF neurons displayed a distribution of BDNF throughout the cell bodies and distal arms of the neurons, while the met-BDNF neurons displayed a distribution of BDNF confined largely to the cell bodies and not extending out to the arms of the neurons. At the cellular-hippocampal level, the researchers measured in vivo hippocampal levels of n-acetyl-asparate in humans, which is a marker of neuronal activity. The Val/Met research participants had significantly lower neuronal activity than the Val/Val research participants.[11] In the following study, the researchers utilized functional magnetic resonance imaging (fMRI) to investigate the role of the Val66Met SNP at the hippocampal and memory levels (Hariri et al. 2003). In this study (a top-down activation experiment), human research participants were first exposed to six images of indoor and outdoor scenes (the encoding phase). Then, later, the research participants were exposed to six additional images of indoor and outdoor scenes and challenged to determine whether or not these new scenes were seen earlier (the retrieval phase). Weinberger and his team found several interesting results from this experiment. First, the Val/Val participants had greater hippocampal activity during both the encoding and retrieval phases of the task in comparison to the Val/Met participants. Second, the Val/Val participants were significantly more accurate at discerning the scenes that they had seen before in comparison to the Val/Met participants. Third, the differences identified were specific to activity in the hippocampus; they did not affect other brain regions. Finally, the results pointed particularly to the role of BDNF (and the impact of the Val66Met SNP) during the encoding phase of the memory task. The following year, Weinberger's team investigated whether or not Val66Met translated into differences in brain morphology, and they discovered that Val/Val participants had larger hippocampal gray-matter volume compared to the Val/Met and Met/Met participants (Pezawas et al. 2004). In the years since Weinberger's publications, research on Val-66Met has continued to elucidate the relationship between that SNP and hippocampal-dependent memory (for a review, see Bath and Lee 2006; but see also Karnik et al. 2010 for a failure to replicate the Weinberger results).

Weinberger's research, in many senses, was quite similar to the research on the levels of mechanism for spatial memory discussed earlier. Like Levine and Nabeshima, that is, Weinberger investigated how differences in the BDNF protein translated into differences at higher levels—differences in molecular distribution, differences in cellular activity, differences in hippocampal activity, and differences in memory. But here is the key difference between Weinberger's research and that of his predecessors—Levine and Nabeshima artificially induced differences in the BDNF protein via experimental intervention, whereas Weinberger found a naturally occurring, actual difference in the BDNF protein due to the Val66Met SNP. And that actual difference accounted for actual variation at the higher levels of mechanism. Remember Waters's point: "in order for there to be a difference maker there must be a difference. And in order for there to be an *actual* difference maker, there must be an *actual* difference" (Waters 2007, 566). What Weinberger's team focused upon (and which his predecessors did not) was the fact that there is naturally occurring, actual variation in the phenomenon under investigation. And that meant there had to be an actual difference (or, more accurately, actual differences) that accounted for that variation. An actual difference was the difference in the Val66Met SNP. In short, Weinberger's attention to the causes of variation responsible for actual variation in memory (as opposed to the typological memory held by the Platonic form of the rat or the Platonic form of the human) brought population thinking to bear on the thinking about the mechanisms of memory. Rather than elucidating the mechanisms responsible for memory, Weinberger and his team were elucidating what I have called elsewhere the "difference mechanisms" responsible for differences in memory (Tabery 2009a).[12]

The Explanatory Bridge: An Exercise in Integrative Pluralism

I introduced the research on BDNF because it shows how the elucidation of the causal mechanisms responsible for the developmental process and the search for the causes of variation responsible for variation in a population can come together to form population thinking about mechanisms. Now here's the key: population thinking about mechanisms is the explanatory bridge between the variation-partitioning and the mechanism-elucidation approaches. I draw on the BDNF example below to show how, explicating the relationship between their (a) things to be explained, (b) causal questions, (c) things that do the explaining, and (d) methodologies (see table 5.1).

Table 5.1
The Components of the Explanatory Divide

	Variation-partitioning approach	Mechanism-elucidation approach
Thing to be explained	Variation in a population	Developmental process
Causal question	How much?	How?
Thing that does the explaining	Cause of variation	Causal mechanism
Methodology	Statistical	Interventionist

(a) *Thing to be explained* Variation in a population (the thing to be explained for the variation-partitioning approach) is the result of actual difference makers in the causal mechanisms of the developmental process (the thing to be explained for the mechanism-elucidation approach). Variation in spatial memory, for example, is the result of actual difference makers in the causal mechanisms responsible for the development of spatial memory.

(b) *Causal question* How-much questions (asked by the variation-partitioning approach) are about how much of the actual variation in the population is accounted for by the actual difference makers in the population, while how questions (asked by the mechanism-elucidation approach) are about how difference makers make their difference in the causal mechanisms. Some of the research on BDNF (i.e., the research by Levine, Nabeshima, and their teams) was designed to elucidate how BDNF made a difference in the development of spatial memory, while other research on BDNF (i.e., the research by Weinberger and his team) was designed to assess how much of the variation was accounted for by the actual difference in the Val66Met SNP.

(c) *Thing that does the explaining* A cause of variation (the thing that does the explaining for the variation-partitioning approach) is an actual difference maker in the causal mechanisms of the developmental process (the thing that does the explaining for the mechanism-elucidation approach). The *BDNF* gene, for example, is an actual difference maker in the causal mechanisms responsible for the development of spatial memory.

(d) *Methodology* Statistical methodologies (utilized by the variation-partitioning approach), such as the analysis of variance, measure how much of the actual variation in the population is accounted for by actual difference makers in the population. Interventionist methodologies (utilized by the mechanism-elucidation approach), such as stimulation or interference experiments, determine how the difference makers make their difference in the causal mechanisms of the developmental process. Some of the research on BDNF (i.e., the research by Levine, Nabeshima, and their teams) utilized interventionist methodologies to elucidate how BDNF made a difference in the development of spatial memory, while other research on BDNF (i.e., the research by Weinberger and his team) utilized statistical methodologies to measure how much of the actual variation in the population of research participants was accounted for by the actual difference in the Val66Met SNP.

Notice that my account of the things to be explained, causal questions, things that do the explaining, and methodologies explicated above makes the line between the variation-partitioning and mechanism-elucidation approaches rather blurry. But in attempting to bridge the two, that is precisely the point and goal. The elucidation of the causal mechanisms responsible for the developmental process and the search for the causes of variation responsible for variation in a population are not necessarily exclusive enterprises; they co-inform one another at the intersection of population thinking about mechanisms. I emphasize the research on BDNF (the BDNF protein, the *BDNF* gene, and the Val66Met SNP) because I take it to be a demonstrative example of scientists bridging the explanatory divide between the two approaches by way of population thinking about mechanisms. So I am not just saying that bridging the explanatory divide is something that can be done; I am saying that it is being done. It is being done by scientists like Weinberger and his team, who engage in population thinking about mechanisms, and generally by a scientific discipline, which engages in the process of moving back and forth between first identifying actual difference makers and then elucidating how those difference makers make their difference.[13]

The Bridge Metaphor

Population thinking about mechanisms is what I have been calling thus far the "explanatory bridge" between the variation-partitioning and mechanism-elucidation approaches. I take the bridge metaphor seriously here. A bridge is a means of spanning two otherwise isolated locations. It is not a means of reducing one location to the other. Nor is it required to be used if staying at one location is sufficient for one's purposes. Likewise, population thinking about mechanisms does not reduce causes of variation to causal mechanisms or reduce how questions to how-much questions. Nor is population thinking about mechanisms required of members of either the variation-partitioning or mechanism-elucidation approaches. If a scientist (say, a geneticist) is only interested in manipulating the actual variation in a trait (say, eye color in fruit flies) by way of manipulating the actual difference makers for that trait, then she is not at all required to cross the explanatory bridge in order to elucidate the causal mechanisms that link the actual difference maker and the actual difference. Likewise, if a scientist (say, a neuroscientist) is only interested in how a difference maker makes its difference in a phenomenon (say, spatial memory) by way of experimentally manipulating that difference maker, then he is not at all required to cross the explanatory bridge in order to determine whether that difference maker is a naturally occurring, actual difference maker. But the bridge is there. If the geneticist wants to elucidate the causal mechanisms that link the actual difference maker to the actual difference or the neuroscientist wants to determine which of the potential difference makers are actual difference makers, then the bridge is available. Population thinking about mechanisms is one type of what Sandra Mitchell calls "integrative pluralism," wherein the answers and results from one discipline are able to co-inform the questions and investigations of another discipline (Mitchell 2003, 2009).

Just as importantly: choosing not to cross the explanatory bridge has consequences as well, for certain questions about intervention and explanation can only be answered from one side of the bridge. If the geneticist, for example, only studies how the actual variation in a trait is manipulated by the actual difference maker for that trait and ignores the causal mechanisms that link the two, then she cannot claim to have identified the only means of intervening to bring about change in the trait. There may be potential difference makers in the uninvestigated causal mechanisms that are also potential sources of intervention. Likewise, if the neuroscientist only studies how a difference maker makes its difference and ignores whether or not the difference maker is a naturally occurring, actual difference maker, then he cannot claim to have identified the source of actual differences in the trait under investigation. He may only have elucidated how a potential difference maker makes its difference. I will return to this point about the limitations associated with staying on one side of the bridge in the next chapter.

To be clear, I do not take it to be the case that bridging the explanatory divide is the same thing as generating a complete scientific explanation of a phenomenon. Scientists like Weinberger and his team have actively bridged the explanatory divide by drawing on the resources of both the variation-partitioning and the mechanism-elucidation approaches to explanation and applied them to the phenomenon of spatial memory. But that does not imply that they have completely explained the phenomenon (as I'm sure they would admit). Far from it. A complete scientific explanation of spatial memory would involve (a) the identification of both the actual and the potential difference makers in the causal mechanisms responsible for spatial memory; (b) the distribution of the actual and potential difference makers in a population; and (c) the elucidation of how those difference makers make or would make their difference in the causal mechanisms of spatial memory (Griffiths and Tabery 2013). Kenneth F. Schaffner (forthcoming, chapters 3 and 4) has conveyed how frustratingly difficult it has been for behavioral geneticists to provide such explanations for even very simple behaviors in very simple organisms (like foraging behavior in the nematode worm, *C. elegans*). The lessons of that research are a variety of rules about how complex even the simplest systems are: many genes can affect a single neuron, many neurons can affect a single behavior, one gene can affect many neurons, and one neuron can affect many behaviors. So if this is the case for a simple behavior in a simple organism, we should expect at least equal complexity when something as complex as spatial memory is considered in mammals. The point of this chapter is that both the variation-partitioning and the mechanism-elucidation approaches can and have contributed to the efforts at generating an explanation of spatial memory, even if that explanation is still far from complete. The result, as Schaffner described it, are "creeping" or "partial" explanations, explanations that link up a difference maker at one level to a difference at another level, even if that difference does not fully account for all the variation in the phenomenon of interest (Schaffner forthcoming, chapter 5).[14]

Scaling the Explanatory Wall

I have argued so far in this chapter that population thinking about mechanisms is the explanatory bridge between the variation-partitioning and mechanism-elucidation approaches. But remember that a number of authors have questioned the existence of such a bridge. Gottlieb, Moore, and Fox Keller all challenged the idea that correlational studies of populations could shed any light on the causal mechanical understanding of individual development (see Perbal 2013 for a similar point). If I am to claim that I have bridged the explanatory divide, then I need to say explicitly how I have scaled the explanatory wall that these authors have constructed.

Exclusive critics of the variation-partitioning approach to studying nature and nurture have drawn a clear line between the partitioners of variation who are stuck studying correlations with statistics and the elucidaters of mechanisms who are instead studying causation with interventionist experiments; "correlation is not causation," the famous saying goes. But a reflection back on the BDNF research suggests that no such clear line exists. Take the research by Weinberger and his team: in their first publication, they focused on the impact of the Val66Met SNP on the molecular and cellular levels; in the second publication, they focused on the impact of the Val66Met SNP on the hippocampal and memory levels. In both cases, they utilized populations (of rats, of humans) to study the mechanisms of memory. And in both cases, they investigated the relationship between an actual difference maker and the actual difference it made. But in one study they investigated how the Val66Met SNP made its actual difference, while in the other study they investigated how much of the actual variation in memory was accounted for by the actual difference maker. Or, consider the broader history of research on BDNF. What we found there was an iterative, co-informational process, wherein scientists from a number of disciplines moved back and forth between first identifying difference makers and then elucidating how those difference makers made their difference. In the 1980s and early 1990s, scientists identified the BDNF protein and then identified the *BDNF* gene. Throughout the 1990s, scientists elucidated how BDNF made its difference in the levels of mechanism responsible for memory. In the 2000s, scientists identified the Val66Met SNP as an actual difference maker. And since 2003, scientists have been investigating how that actual difference maker makes its actual difference.

The challenge for critics of the variation-partitioning approach is to identify where in this history of BDNF research correlation ends and causation begins. As I see it, when the individual studies are treated in isolation, they're *all* correlational—even the paradigmatic examples of mechanism-elucidation via interventionist experiment. Levine and his colleagues linked up BDNF to the NMDA receptor via the TrkB receptor, but all they showed was that the BDNF binding increased the probability of the NMDA receptor being open. Weinberger and his colleagues linked up the Val66Met SNP to memory, but all they showed was that there were statistically significant differences between the Val/Met and the Val/Val research participants. In sum, a causal picture of the mechanism emerges, not from any individual study, but from a combination of the results of many studies, with multiple correlations at multiple levels all aligning to track how a difference at one level translates into differences at other levels.

The general problem that these exclusive attempts share is confusion between an inability to complete a causal mechanical explanation and an inability to contribute to a causal mechanical explanation. Gottlieb, Moore, and Fox Keller are certainly correct in saying that studying the causes of variation responsible for variation in a population will not complete the goal of elucidating the causal mechanisms of the developmental process. But what single study could? At least in the case of BDNF, the research on causes of variation responsible for variation in memory did contribute to the process. The early twin, adoption, and family studies pointed out that the genome as a whole was a cause of variation and supported looking for a more specific difference maker at the molecular genetic level (i.e., the *BDNF* gene). The fMRI study linked the actual variation in the Val66Met SNP of the *BDNF* gene up to actual variation in hippocampal activity during hippocampal-related memory and supported looking more closely at the encoding phase of the causal mechanism. These studies of causes of variation in populations, that is, did shed light on and did provide understanding of the mechanical causes of development.

Moore, in considering an earlier formulation of the explanatory bridge I am offering here, admitted, "such a strategy *would* permit the integration of approaches that Tabery (2007b) has called for, because it could generate information that would explain both the appearance of characteristics in individuals and differences in those characteristics across members of a population" (author's emphasis, Moore 2008a, 377). However, he then countered,

> adopting this strategy would still leave the extant literature of quantitative behavior genetics explanatorily isolated from literatures on the mechanical causes of development; just because it is possible to identify developmental factors that contribute to differences in phenotype does not mean that quantitative behavior geneticists—who for the past century have used correlational approaches to account for variation across populations—have done any such thing. (ibid., 377–378)

Moore here is making a point similar to the one I made at the end of the "population thinking" section earlier—that members of the variation-partitioning approach, such as human behavioral geneticists, historically have often ignored or black-boxed the causal mechanisms linking actual difference makers to the actual differences they made. But that historical trend should not be confused with an epistemological necessity. Moore, then, admits of the possibility of the explanatory bridge between the search for the causes of variation responsible for variation in a population and the elucidation of the causal mechanisms responsible for the developmental process. The question becomes one of whether or not (or how often) the bridge is actually utilized by members of the variation-partitioning approach. Now, I certainly cannot say that all members of the variation-partitioning approach (be they historical members like Fisher or contemporary members like practicing behavioral geneticists) attempt to bridge their research with research on causal mechanisms. But I can say that many do, as the examples from this chapter attest. Weinberger and his team are doing it. And Kendler explicitly framed a vision for how the epidemiological paradigms of psychiatric genetics can integrate with the molecular paradigms. Thus, the lesson is that while searchers for causes of variation responsible for variation in a population may have historically stayed on one side of the bridge and ignored the causal mechanisms of the developmental process, this historical trend is abating as more and more members of the variation-partitioning approach are crossing the bridge to undertake population thinking about mechanisms. If anything, advocates of the mechanism-elucidation approach should welcome this trend.

What about Fox Keller's concern regarding Waters's actual difference maker concept? Remember, after emphasizing distinctions between traits and trait differences and how questions and how-much questions, she worried:

> With Waters's interpretation of causal statements of the form "x causes y" as meaning "a difference in x causes a difference in y," it becomes all but impossible to mark, or even to recognize, the distinction between the two questions, and correspondingly easy to think that by answering the second [how-much question], one has also answered the first [how question]. (Fox Keller 2010, 39)

The short answer here is: precisely, and that's Waters's achievement. The long answer is: Fox Keller is absolutely right that answering a how-much question (about how much of an actual difference an actual difference maker makes) is not the same thing as answering a how question (about how the actual difference maker makes its difference). And, indeed, great care needs to be taken to mark and recognize the distinction. But a linguistic distinction does not necessarily equate to an explanatory divide. We can (and should) keep how-much and how questions distinct, and yet simultaneously see how, in the case of the research on BDNF, answers to the one can and did co-inform answers to the other, even though answers to the one were not simultaneously answers to the other.

Longino presents a slightly different explanatory wall. Although she pluralistically welcomes contributions from the variation-partitioning approach to understanding human behavior, recall that her pluralism is non-eliminative in nature because the different disciplinary approaches that she considers investigate different portions of the causal space of human behavior. As I have suggested by way of the BDNF example, however, different investigations of different portions of the causal space can still be brought together to co-inform one another. The population thinking about mechanisms undertaken by Weinberger and his team built off of quantitative behavioral genetics, molecular behavioral genetics, and neuroscience.

Longino, like Moore, also considered an earlier formulation of the integrative pluralism I am offering here. And in response, she surmised, "To the extent that Tabery's is a form of pluralism, it is a moderate pluralism too weak to accommodate the incongruencies of causal spaces investigated by the different approaches" (Longino 2013, 145). My version of integrative pluralism via population thinking about mechanisms is indeed a "moderate pluralism" when judged against Longino's project. But "moderate" is a relative term, and this is where my earlier point about our different scopes is relevant. When considering the wide net that Longino casts (research on human behavior generally), my explanatory bridge between the variation-partitioning and mechanism-elucidation approaches is indeed moderate in its scope. For example, Longino points out that some disciplines investigate the distribution or frequency of a particular phenomenon like crime; such investigations of patterns are neither about the development of criminality nor about the causes of variation responsible for variation in criminality (Longino 2013, 116–121). And so my explanatory bridge is indeed moderate when judged against Longino's scope, in that population thinking about mechanisms has nothing at all to say about such investigations. But that is not the scope against which to judge my effort. My scope is the interaction debates and the members of the two approaches who have repeatedly faced off over the last century. The persistence of those debates and the exclusive analyses by the likes of Gottlieb, Moore, and Fox Keller should convey the fact that integratively bridging the explanatory divide between the variation-partitioning and mechanism-elucidation approaches is a significant achievement, even if it is "moderate" by another scope.

It is not moderate, for example, when judged against the scope of Kendler's paradigms. Kendler, recall, sought to integrate basic and advanced genetic epidemiology (paradigms 1 and 2) with gene-finding and molecular genetics (paradigms 3 and 4). But he warned that such an integration will be "murky." The history of research on BDNF, however, shows just how these paradigms can be integrated via population thinking about mechanisms. The first two paradigms identified actual difference makers, albeit at a very general level (e.g., the entire genome). The third paradigm pinpointed a more specific actual difference maker in the form of the *BDNF* gene. And then the fourth paradigm, in coordination with research undertaken by neuroscientists, has focused on understanding how the *BDNF* gene made its actual difference by tracing the actual difference up through the levels of mechanism responsible for one type of memory. Though the result may not be a complete explanation of spatial memory, it is a "creeping" or "partial" explanation, and that is an achievement even still (Schaffner forthcoming, chapter 5).

Back to the Interaction of Nature and Nurture

This chapter was devoted to bridging the explanatory divide between the variation-partitioning and mechanism-elucidation approaches. More specifically, it was devoted to integrating the first four components of those traditions: (a) their different things to be explained, (b) their different causal questions, (c) their different things that do the explaining, and (d) their different methodologies. I argued that population thinking about mechanisms, which unites the resources from the philosophy of mechanisms with Waters's actual difference maker concept, was that explanatory bridge.

That still leaves, however, the last component of the explanatory divide—the variation-partitioning and mechanism-elucidation approaches' very different concepts of interaction. Remember, it was the concept of interaction (or, rather, the history of the debates over that concept) that brought us here in the first place. The members of the two approaches, from the origin*(s)* of the concept all the way to the twenty-first-century debate over the serotonin transporter gene, have had very different answers to the following questions: what is interaction? Why and how should interaction be investigated? And what is the empirical evidence for interaction? In the next chapter I will draw on the explanatory bridge developed here to provide an integrated concept of interaction. That integrated concept will then allow for answering these questions from a perspective that neither the members of the variation-partitioning approach nor the members of the mechanism-elucidation approach have considered.

6 The Interaction of Nature and Nurture: An Integrated Concept

Fisher and Hogben, Jensen and Lewontin, Moffitt and Caspi, alongside their defenders and critics—for close to a century now, disputants at each turn of the interaction debates faced off over the following three questions:

- the conceptual question—what is interaction?
- the investigative question—why and how should interaction be investigated?
- the evidential question—what is the empirical evidence for interaction?

In the previous chapter, I intentionally avoided any mention of the interaction of nature and nurture because the issues there were about the nature/nurture debate generally, not about the more specific interaction debates. The goal of that chapter was to construct an explanatory bridge between the variation-partitioning and mechanism-elucidation approaches in terms of the thing to be explained, the causal question about that thing, the thing that does the explaining, and the methodology used to provide that explanation. Population thinking about mechanisms offered that explanatory bridge, presenting a relationship wherein the answers and results from one approach could co-inform the questions and investigations of the other approach. With that general relationship in place, it is time to return to the conceptual, the investigative, and the evidential questions. I will proceed by treating each question in turn.

The Conceptual Question: What Is Interaction?

The conceptual, investigative, and evidential questions all arose time and again in the interaction debates. But there is an imbalance in terms of the weight of each of these questions, in that the investigative and the evidential questions presume an answer to the conceptual question. Questions about the way to investigate interaction or the empirical evidence for interaction presuppose having already answered what the investigation is investigating and what the evidence is evidence of. As a result, we must start with the conceptual question.

Fisher had a very clear answer to this question. Interaction was a statistical measure of the breakdown in additivity between the main effects of nature and nurture. This is what I called the "biometric concept of interaction." Fisher, when he first worried about the problem, paved the way for all future variation-partitioning answers to the conceptual question—interaction was a "deviation from summation formula" and the "non-linear interaction of environment and heredity" (Fisher and Mackenzie 1923). For Fisher, interaction was conceptualized as an absence—an absence of summation, an absence of linearity, an absence of additivity.

As we saw in chapters 3 and 4, Fisher's answer to the conceptual question was carried forward by subsequent members of the variation-partitioning approach to nature and nurture as well as philosophical defenders of that approach; and, importantly, these subsequent members/defenders were particularly critical of any discussions of interaction that conceptualized it differently. Jay Lush specifically considered (but then dismissed) the "nonadditive combination effects of heredity and environment" (Lush 1937, 64), and Jensen bemoaned the "considerable confusion concerning the meaning of interaction" (Jensen 1969, 39). On the philosophy side, Neven Sesardic distinguished statistical interaction from commonsense interaction and similarly complained about confusions of the two (Sesardic 2005). And in the most recent episode of the interaction debates, critics of Moffitt and Caspi warned against "misconceptions about gene-environment interactions"; the appropriate conception of interaction, these authors clarified, "always refers to a measure of statistical interaction" (Zammit, Owen, and Lewis 2010, 65).

Hogben answered the conceptual question quite differently. Interaction was not simply the absence of perfect additivity between the hereditary and environmental causes of variation; interaction was the presence of a unique source of variation—what Hogben called the "third class of variability," which "arises from the combination of a particular hereditary constitution with a particular kind of environment" (Hogben 1932, 98). This is what I called the "developmental concept of interaction," because, for Hogben, it was a uniquely developmental phenomenon.

As we also saw in chapters 3 and 4, Hogben's answer to the conceptual question was carried forward by subsequent proponents of the mechanism-elucidation approach to nature and nurture as well as philosophical defenders of it. Waddington defined interaction as "a difference of degree in environmental sensitivity to the development controlled by two genotypes" (Waddington 1957, 94). And during the IQ controversy, Lewontin and Layzer both made much of interaction, emphasizing the "complicated developmental process in which genetic and environmental factors are inextricably mingled" (Layzer 1972, 275). Subsequently, a host of Lewontinians carried this concept of interaction into the history and philosophy of biology. And in the most recent episode of the interaction debates, Michael Rutter explained how his (as well as Moffitt and Caspi's) concept of interaction followed from Hogben's formulation (Rutter 2008).

Interaction as Interdependent Actual Difference Makers

Building off the vision of population thinking about mechanisms introduced in chapter 5, my answer to the conceptual question looks like this: *interaction is the interdependence of actual difference makers in the causal mechanisms responsible for a phenomenon*. Population thinking about mechanisms, recall, embedded Waters' concept of an actual difference maker in the philosophy of mechanisms. According to typological thinking about mechanisms, a scientist explains a phenomenon by identifying and manipulating the variables in the mechanisms responsible for that phenomenon, thereby determining how those variables are situated in and make a difference in the mechanism. According to population thinking about mechanisms, a scientist explains variation in a phenomenon by identifying and manipulating the actual difference makers in the mechanisms responsible for that phenomenon (the "difference mechanisms"), thereby determining how those actual difference makers are situated in and make actual differences in the difference mechanism.

An example will help convey this notion of interaction as interdependent actual difference makers. In chapter 5, I used the example of research on brain-derived neurotrophic factor (BDNF) to show how this population thinking about mechanisms works. Turning to interaction, however, we need an example with more than one actual difference maker in the mechanisms responsible for the phenomenon; interaction results from interdependent actual difference makers, and so there must be more than one actual difference maker in order for there to be interdependence among them. The serotonin transporter gene example, which took center stage in chapter 4, will work perfectly in this regard.

Recall from chapter 4 that the serotonin transporter gene example first became noteworthy in the interaction debates when Moffitt and Caspi published their 2003 study identifying a case of gene-environment interaction between the serotonin transporter gene, exposure to stressful life events, and the development of clinical depression (Caspi et al. 2003). They found that individuals with the short form of the serotonin transporter gene were more vulnerable to developing depression when exposed to stressful life events. This original result was subjected to dozens of replications and multiple meta-analyses, and the evidence was a mixed bag of positive replications/meta-analyses and negative replications/meta-analyses. I will return to the evidence for/against the serotonin transporter gene study below. For the time being, though, let's imagine Moffitt and Caspi found a legitimate case of interaction between the serotonin transporter gene, stress, and depression. Doing so will allow me to both (a) now provide an example of what it means to understand interaction as interdependent actual difference makers and (b) later hold up that example as the measure by which evidence for or against interaction's existence should be judged.

The Serotonin Transporter Gene and Stressful Life Events: Interdependent Actual Difference Makers in the Causal Mechanisms Responsible for Depression

Why do some people develop clinical depression, while others do not? Throughout the 1970s and 1980s, a variety of epidemiological research suggested that both genes and the environment contributed to the risk of developing depression. On the environmental side, for instance, research by George Brown at Bedford College, London, and by Eugene Paykel at Yale University and then St. George's Hospital Medical School, also in London, identified stressful life events (such as divorce, death of a loved one, and serious illness) as contributors to depression (for reviews, see Paykel 1978 and Brown 1987). On the genetic side, a number of twin and adoption studies from human behavioral geneticists implicated a genetic component in depression and associated traits, such as neuroticism (Meltzer and Arora 1988; Loehlin 1989; for reviews, see Lander and Schork 1994; Plomin, Owen, and McGuffin 1994). These epidemiological studies suggested stressful life events and something genetic acted as causes of variation responsible for variation in depression. But these studies did not elucidate how these causes of variation made their difference. How does stress "get in the brain," as it were? And how do genes actually contribute to this process?

Answering these "how" questions involves the story of serotonin. In the 1940s, serotonin (5-hydroxytryptamine, or 5-HT) was discovered independently by Vittorio Erspamer at the University of Pavia and the University of Rome, and Irvine Page at the Cleveland Clinic; Page also pinpointed serotonin as acting in the brain (Rappaport, Green, and Page 1948; Erspamer and Asero 1952; Twarog and Page 1953; see Whitaker-Azmitia 1999 for a history of this discovery). In the 1950s, Dilworth Woolley and Elliott Shaw suggested there was a link between serotonin and mental illness because substances that acted on the serotonergic system, such as lysergic acid diethylamide (LSD), generated mental disturbance (Woolley and Shaw 1954; Shaw and Woolley 1956). Woolley and Shaw's research implicated serotonin as a potential difference maker that could be artificially manipulated to bring about changes in mental condition. Two decades of pharmacological R&D followed, generating numerous drugs designed to act on serotonin levels in order to treat conditions ranging from hypertension and Parkinson's disease to depression and eating disorders (Sjoerdsma and Palfreyman 1990). Many of these drugs, it became clear, affected serotonin levels by acting on the serotonin transporter (5-HTT; Raisman, Briley, and Langer 1979). Structurally, the serotonin transporter is a protein that resides on the membranes of neuronal synapses. Functionally, during neurotransmission, when serotonin floods into the synapse between the presynaptic and the post-synaptic neuron, the serotonin transporter reabsorbs serotonin back into the presynaptic neuron, thus terminating serotonergic neurotransmission; in this way, the serotonin transporter regulates both the magnitude and the duration of serotonergic neurotransmission (Kanner and Schuldiner 1987). Serotonin was long known to play an important role in depression (Meltzer 1990); in the early 1980s, the serotonin transporter in particular was implicated in this relationship, with reduced serotonin transporter activity linked to increased risk of depression (Briley et al. 1980; Langer et al. 1981).

With epidemiological research suggesting that something genetic was contributing to depression risk and molecular research suggesting that somehow the serotonin transporter was contributing to depression risk, the next question became: what is the genetic basis of the serotonin transporter, such that actual differences in it account for actual differences in depression risk? Klaus-Peter Lesch spent much of the early- to mid-1990s answering precisely this question. Lesch is a clinical psychiatrist at the University of Würzburg in Germany. He and his colleagues first identified and localized the gene responsible for producing the serotonin transporter—*SLC6A4* (Lesch et al. 1994). The gene contains 14 exons and spans approximately 31 kilobases on chromosome 17 in humans (see also Ramamoorthy et al. 1993; Gelernter, Pakstis, and Kidd 1995). With the gene for the serotonin transporter identified, Lesch next tracked down the source of variation in that gene that would account for variation in the serotonin transporter product. Upstream of the main coding region, Lesch found a polymorphism in the promoter region—*5-HTTLPR* (for serotonin transporter-linked polymorphic region; Heils et al. 1995, 1996). *5-HTTLPR* can be relatively short ("s") or long ("l"), and individuals get one copy of this gene from each of their parents, so the possibilities are s/s, s/l, and l/l. This physical difference in length translates into a functional difference—the short form is less efficient at producing serotonin transporter proteins than the long form. Finally, Lesch linked variation in *5-HTTLPR* up to variation in depression. In a study of 505 subjects (19% s/s, 49% s/l, and 32% l/l), Lesch and his colleagues reported a number of striking differences between individuals carrying the long form of *5-HTTLPR* and individuals carrying the short form (Lesch et al. 1996; see also Collier et al. 1996). At the molecular level, cells with the l/l form were twice as efficient at serotonin uptake as cells with either the s/s or s/l forms. A number of personality tests were then used to assess the behavioral level, and Lesch focused particularly on measures of neuroticism because of that trait's strong association with anxiety-related psychiatric disorders like depression. Individuals with either the s/s or s/l forms, it turned out, scored significantly higher on measures of neuroticism than individuals with the l/l form.

Like Daniel Weinberger's research on BDNF (discussed in chapter 5), Lesch's research on the serotonin transporter gene was designed to first identify and then explicate how an actual difference maker makes its difference. He was investigating how differences at lower levels percolate up to differences at higher levels. That actual difference maker (the promoter region of the serotonin transporter gene) directly affects the efficiency of serotonergic neurotransmission, and that efficiency in turn results in behavioral differences. But just how that molecular level links up to the behavioral level remained unclear. Several studies that appeared in the early 2000s made this connection; and, in fact, Weinberger was central to this story too. Weinberger focused his attention on the amygdala. The amygdala are two small, almond-shaped clusters of nuclei deep in the brain. Among other things, the amygdala figures centrally in emotional learning and emotional memory; for instance, fear conditioning (the process wherein organisms learn to associate fearful stimuli with negative emotion) runs directly through the amygdala. Weinberger's team utilized neuroimaging to assess whether differences in the serotonin transporter gene translated into differences in brain function that could be associated with the higher-level behavioral differences. In the first study, research participants were divided up based on the serotonin transporter gene variant they carried (s/s vs. s/l vs. l/l); they were then exposed to fearful stimuli (angry/frightened faces) while their brains were neuroimaged. The research participants with either the s/s or s/l variants, it turned out, experienced much greater amygdala activity during this exercise than the participants with the l/l variant (Hariri et al. 2002). In a second study, Weinberger's team tracked differences in the serotonin transporter gene to differences in brain morphology and functional connectivity between the amygdala and other regions of the brain such as the cingulate, which also figures into emotional learning and memory. S-carriers had smaller cingulate and amygdala volume than l/l-carriers; s-carriers also had poorer amygdala-cingulate functional connectivity than l/l-carriers (Pezawas et al. 2005). Weinberger's research, in short, revealed how differences in the serotonin transporter gene linked up to higher-level behavioral differences by way of differences in amygdala structure and function at the brain-system level.

Although Lesch did associate differences in the serotonin transporter gene with differences in anxiety-related personality traits, it should be emphasized that the difference at the genetic level accounted for a very small portion of the total variation at the behavioral level—only 3% to 4% (Lesch et al. 1996). This suggested that the serotonin transporter gene was by no means *the* actual difference maker, making it a very poor predictor of clinical depression when taken by itself. The innovative move of Moffitt and Caspi was to combine this difference maker from the genetic side with a known actual difference maker from the environmental side—stress. On the genetic side, Moffitt and Caspi divided up the Dunedin Study research participants ($n = 847$) based on the serotonin transporter gene variant they carried: s/s (17%), s/l (51%) and l/l (31%). On the environmental side, they divided up the research participants based on the number of stressful life events they had encountered so far in life: none (30%), one event (25%), two events (20%), three events (11%), and four or more events (15%). The combined genetic and environmental difference makers made for a much stronger predictor of depression than either of the difference makers alone. Individuals with the s/s variant were much more susceptible to developing depression when exposed to stressful life events compared to individuals with the l/l variant; indeed, individuals with the s/s variant who were exposed to four or more stressful life events had approximately a 40% probability of experiencing an episode of major depression, and individuals with the s/s variant who were exposed to childhood maltreatment had approximately a 60% probability of experiencing an episode of major depression (Caspi et al. 2003).[1]

I said earlier that, according to population thinking about mechanisms, a scientist explains variation in a phenomenon by identifying and manipulating the actual difference makers in the mechanisms responsible for that phenomenon (the "difference mechanism"), thereby determining how those actual difference makers are situated in and make actual differences in the difference mechanism. By 2010, Moffitt and Caspi had the above history that led up to their 2003 study as well as seven years of research following their 2003 study to reflect upon. With that research in mind, they offered the following "unifying mechanism":

> In theory, *5-HTTLPR* S-carriers are characterized by the stable trait of negative affectivity [or neuroticism] that is converted to psychopathology only under conditions of stress; just as glass is always characterized by the trait of brittleness but shatters only when a stone is thrown. Negative affectivity represents the potential for excitability of anxiety and fear neural circuits, and is characterized by an attentional bias toward negatively valenced information and a cognitive sensitivity to perceive threat. (Caspi et al. 2010, 515)

Let me translate this into population thinking about mechanisms. The serotonin transporter gene is an actual difference maker for neuroticism. More specifically, at the genetic level there are actual differences in the promoter region of the serotonin transporter gene (*5-HTTLPR*); these genetic differences lead to actual differences at the molecular level in terms of transcriptional efficiency; molecular differences lead to actual differences at the cellular level in terms of serotonergic neurotransmission; cellular differences lead to actual differences at the brain systems level in terms of amygdala morphology and function; and amygdala differences lead to actual differences at the behavioral level in terms of neuroticism. That neuroticism, however, is only translated into a clinical disorder like depression if it is exposed to a second actual difference maker—stress. And it is in this sense that interaction is the interdependence of actual difference makers in the causal mechanisms responsible for a phenomenon. In environments without stressful life events, individuals with the s-variant are more susceptible to neuroticism than individuals with the l/l variant, but they are no more susceptible to depression. When it comes to depression, it takes exposure to the environmental difference maker to actualize the genetic difference maker. The one actual difference maker moderates the effect of the other actual difference maker and, in so doing, the actual difference makers in the mechanism are interdependent.

With this answer to the conceptual question in place, let's return to the original biometric and developmental concepts of interaction. For practitioners of the variation-partitioning approach to nature and nurture, interaction was a statistical phenomenon defined by an absence—a statistical measure of the breakdown in additivity between the main effects of nature and nurture. For practitioners of the mechanism-elucidation approach, interaction was a developmental phenomenon defined by a presence—variation that resulted from differences in unique, developmental combinations of nature and nurture. On the account I have offered here, both sides are partially right, but incomplete because they each leave out the other side. Interaction is the interdependence of actual difference makers in the causal mechanisms responsible for a phenomenon. The biometric concept is a statistical measure of a developmental phenomenon. And, because of this, interaction is both an absence and a presence: it is the absence of variation due to independent actual difference makers; it is the presence of variation due to interdependent actual difference makers.

Helen Longino, in her recent book, considers the account of interaction that I offer here and then warns,

> Tabery's solution suggests that the population [i.e. variation-partitioning] and developmental [i.e. mechanism-elucidation] approaches are reconcilable, that each identifies factors relevant to the other, but from a different vantage point. If this solution means, however, that G×E interaction is effective only for the proportion of variance in a population not accounted for by G and E taken separately, it will not satisfy the developmentalists who argue that their form of interaction characterizes every individual. Nor will it satisfy the quantitative behavioral geneticist who argues that the causal processes internal to individuals do not illuminate patterns of variation within a population. (Longino 2013, 145)

Both sides, however, should be satisfied when the following is recognized: interdependent actual difference makers are not the only type of developmental interactions that occur in the causal mechanisms responsible for a phenomenon. Many developmental interactions involve the interdependence of just potential difference makers; these difference makers are crucial parts of the causal mechanisms, but they generate no actual variation. Likewise, many developmental interactions involve a combination of interdependent actual and potential difference makers. Reflecting back on the episodes of the interaction debate from part I, the scientists who elucidated mechanisms made much of the developmental nature of interaction, while the scientists who partitioned variation forcefully replied that interaction was a statistical concept that had nothing to do with development. The correct answer lies in between: interaction is developmental in nature, but it does not encompass all forms of developmental interaction.

The Investigative Question: Why and How Should Interaction Be Investigated?

With the interaction of nature and nurture now understood as the interdependence of actual difference makers in the causal mechanisms responsible for a phenomenon, the investigative question concerning why and how this interaction should be investigated can be addressed. Fisher's answer to the conceptual question shaped his answer to the investigative question. Since interaction was conceptualized as the absence of additivity between the main effects of nature and nurture, interaction was a nuisance to getting on with the task of measuring those relative contributions of nature and nurture. Fisher, however, had a method for assessing how much of a nuisance interaction really was—his analysis of variance. The analysis of variance partitioned sources of variation and attributed them to the assorted causes of variation. If the variation due to interaction was small, it could be ignored; if it was large, it could be eliminated with a transformation of scale.

Fisher's answer to the investigative question was carried forward by the members and defenders of the variation-partitioning approach, who continued to conceptualize interaction just as Fisher did. Jensen also concerned himself with interaction, but only because it would have created a problem for his attempts to measure the relative contributions of nature and nurture to IQ. Jensen, recall, sought to assess the degree to which differences in IQ between racial groups could be accounted for by differences in genotype, and so the heritability of IQ (the proportion of total variation attributable to genetic variation) was his prime concern. Jensen, however, had a method for assessing how much of a nuisance interaction really was—Fisher's analysis of variance (Jensen 1969). If the variation due to interaction was small for IQ, it could be ignored; if it was large, it could be eliminated with a transformation of scale. Sesardic again echoed Jensen's answer: interaction would pose a problem for heritability measurements, but the analysis of variance was perfectly capable of detecting whether or not that nuisance was a reality (Sesardic 2005). Jumping ahead to the serotonin transporter gene debate of the twenty-first century, the focus shifted from whole genotype–environment interaction to candidate gene–environment interaction. And there, critics of Moffitt and Caspi's study on the interaction between the serotonin transporter gene and stressful life events harkened back to the earlier criticisms of interaction. The investigation of main effects (via genome-wide association studies) must take primacy over the investigation of interaction, and it was only when main effects existed that you could then spend valuable research resources to go looking for an interaction effect (Zammit, Owen, and Lewis 2010; Flint and Munafò 2012). Moreover, if interaction did exist, it could be eliminated with a transformation of scale (Eaves 2006).

Hogben answered the conceptual question differently than Fisher, and so it is not surprising that he answered the investigative question differently too. For Hogben, interaction was not a nuisance to be dismissed; it was to be sought out for the light it shed on the developmental process. When it came to seeking out interaction, Hogben was fine with using Fisher's statistics to detect causes of variation, but he simultaneously cautioned against using those same statistics to ascribe degrees of influence to those causes (Hogben 1933a). Once the causes of variation were identified, Hogben counseled moving toward more interventionist methodologies in order to investigate how different genotypes respond to different environments and how those genetic and environmental factors combined during development to generate a phenotype.

Hogben's answer to the investigative question was carried forward by the practitioners and defenders of the mechanism-elucidation approach to nature and nurture. During the IQ controversy, Lewontin and Layzer reiterated Hogben's emphasis on seeking out interaction rather than ignoring or dismissing it "like an uninvited party guest" (Layzer 1972, 471). Lewontin did famously label the analysis of variance "useless" (Lewontin 1974, 410), and a number of historians and philosophers of biology fixated on that verdict and carried it forward as the final word on the issue (see Downes 2009 for a review of Lewontin's influence). But when we look at Lewontin's wider writings, it is more accurate to characterize Lewontin as being in line with Hogben here—critical of partitioning variation as an end point (with heritability estimates), rather than as a first step toward elucidating mechanisms (Simpson, Roe, and Lewontin 1960). When it came to the serotonin transporter gene debate of the twenty-first century, scientists such as Moffitt, Caspi, and Rutter embraced Hogben's position: statistical analyses of populations could indeed be useful for identifying cases of interaction between particular genes and particular environmental factors. That identification, however, was neither a nuisance nor an end in itself; rather, interaction was to be sought out as a route to investigating the causal mechanisms of the developmental process that gave rise to complex human traits (Moffitt, Caspi, and Rutter 2005; Caspi and Moffitt 2006; Rutter 2008).

Investigating Interdependent Actual Difference Makers

With interaction conceptualized as interdependent actual difference makers in the causal mechanisms responsible for a phenomenon, my answer to the investigative question looks like this: *interaction can and should be sought out because it sheds light on the causal mechanisms responsible for a phenomenon.* Statistical analyses of causes of variation in a population contribute to but do not complete such investigations to the extent that they can detect interdependent actual difference makers and thereby direct interventionist methodologies toward elucidating how those difference makers modify one another during the developmental process.

I highlighted the research on BDNF in chapter 5 because I took it to be a demonstrative example of scientists engaged in population thinking about mechanisms. Likewise, I have highlighted the research on the serotonin transporter gene in this chapter because I take it to be a demonstrative example of scientists attempting to investigate interdependent actual difference makers in the causal mechanisms responsible for depression. And as with the BDNF case previously, the serotonin transporter gene case revealed an iterative, co-informational process, wherein scientists moved back and forth between detecting difference makers and elucidating how those difference makers make their difference. For the serotonin transporter gene case, the story began with a combination of interventionist methodologies that identified serotonin as a potential difference maker alongside statistical studies of populations that detected something genetic and stressful life events acting as actual difference makers. Lesch brought these two strands together when he sought out the genetic basis of the serotonin transporter and tracked down how actual differences in the serotonin transporter gene translated into actual differences at the molecular and cellular levels. Weinberger carried this forward by linking up the actual differences at these lower levels with actual differences at the behavioral level by way of actual differences in amygdala structure and function. Moffitt and Caspi combined this research on a genetic actual difference maker with the research on an environmental actual difference maker and ultimately proposed a "unifying mechanism" that pulled the various threads together—a genetic actual difference maker that leads to actual differences in neuroticism regardless of exposure to stressful life events, but only actual differences in depression when exposed to actual differences in stressful life events.[2]

This answer to the investigative question aligns me with psychologist Douglas Wahlsten, who has written extensively on the interaction of nature and nurture. Wahlsten treats the analysis of variance as being "in the service of interactionism," and so he warns against dismissing the analysis of variance as entirely useless (Wahlsten 2000). Likewise, he describes Fisher's statistical method as being able to "interrogate nature" precisely because "the great merit of the [analysis of variance] method is that it can readily detect interaction" (Wahlsten 1990, 110). Wahlsten's influential contributions have been methodological in nature, devoted to addressing the investigative question alone. In contrast, I have derived an answer to the investigative question by way of the more fundamental answer to the conceptual question. Doing so allows me to say why the analysis of variance can "interrogate nature" and be "in the service of interactionism" without completing the task—because the methodology can detect interdependent actual difference makers in the causal mechanisms of development, without being able to determine how the actual difference makers developmentally interact in the causal mechanisms.

Importantly, though, acknowledging the analysis of variance's ability to identify interdependent actual difference makers should not be confused with an endorsement of using the analysis of variance to compute heritability measures. The former contributes to elucidating the causal mechanisms of development; the latter does not. This is a point that psychologist Eric Turkheimer has been making for several decades now. In 1991, with Irving Gottesman, Turkheimer dubbed the "first law of behavioral genetics" to be that all human traits are heritable to some degree (Turkheimer and Gottesman 1991, 411). But he didn't introduce this law as a rallying cry for variation-partitioners to go perform more heritability studies; on the contrary, calling all behavioral traits heritable became Turkheimer's persistent plea for heritability studies to halt, since there was nothing left for them to contribute to the study of nature and nurture other than a reaffirmation of the fact that genes are somehow contributing to complex human traits (Turkheimer 1998, 2000, 2008, 2011). Turkheimer's point is a crippling criticism of much of the history of the variation-partitioning approach to studying nature and nurture, especially as it relates to their dismissals of interaction (see also Sarkar 1998; Kaplan 2000; Taylor 2006, 2010; Lickliter and Honeycutt forthcoming). Proponents of the variation-partitioning approach, from Fisher to present-day behavioral geneticists, time and time again emphasized that interaction had nothing to do with the causal mechanisms of individual development. Because they did not see its developmental nature, they simultaneously did not see the valuable information it offered, and so they opted for either ignoring the phenomenon or making it go away with a transformation of scale. What a waste! Rather than utilizing a result of interaction to then guide a subsequent investigation of the causal mechanisms that would give rise to that interaction effect, too often these scientists sought to eliminate the nuisance in exchange for deriving yet another heritability measure that contributes nothing to the causal mechanical investigation.

Now, members and defenders of the variation-partitioning approach could reply to this criticism by claiming that they are not worried about causal mechanisms; they do not suffer from "mechanism envy." Fair enough, and this goes back to the explanatory bridge from chapter 5—members from each approach need not necessarily cross the bridge, and so partitioners of variation are certainly allowed to stay on their side and ignore the causal mechanisms that link actual difference makers to actual differences. But, as I warned there, staying on one side of the bridge has serious consequences. The consequences pertain to both intervention and extrapolation. If one is interested in either intervening on current populations or environments in order to bring about different phenotypic results, or extrapolating to other possible populations or environments in order to assess what phenotypic results will occur there, then knowledge of the causal mechanisms that link difference makers to differences is crucially important. When it comes to intervention, actual difference makers are not the only route to bringing about different phenotypic results; potential difference makers can be manipulated to alter outcomes too (as research on the serotonergic system revealed). But confined to just the variation-partitioning side of the bridge, these scientists necessarily cut themselves off from potential difference makers in exchange for actual difference makers; thus, they necessarily cut themselves off from a whole class of intervenable difference makers (Griffiths and Tabery 2013). Likewise, when it comes to extrapolation, understanding the causal mechanism allows for predicting how the system will respond to new environmental inputs (Steel 2008, chapter 5; Tabery and Griffiths 2010).

The members and defenders of the variation-partitioning approach have not always been sensitive to these consequences. For instance, Jensen defended his assault on compensatory education's failure from Lewontin's argument from interaction by claiming he himself was focused on the actual, whereas Lewontin was fixated on the possible (Jensen 1970, 1975). Likewise, Sesardic criticized Lewontin and the Lewontinians for their "curious triumph of the *possible* over the *actual*" (author's emphasis, Sesardic 2005, 85). Jensen and Sesardic's distinction between the possible and the actual is fundamentally about extrapolation; confining oneself to the actual foregoes the ability to extrapolate to the possible. But, in judging compensatory education a failure, Jensen himself weighed in on both the actual and the possible; compensatory education was unlikely to succeed according to Jensen because IQ had a high heritability. In making that claim about compensatory education's existing and potential failure, however, Jensen overextended himself, because the variation-partitioning approach upon which he relied simply does not warrant this type of extrapolation.

The Evidential Question: What Is the Empirical Evidence for Interaction?

Like the conceptual and investigative questions, the empirical evidence for interaction has been a subject of dispute for close to a century now. With the evidential question, though, it is especially important to keep the differences between each instantiation of that debate distinct. For Fisher and Hogben in the context of the eugenics controversy, the question was: what is the empirical evidence for heredity-environment interaction in nature? For Jensen and Lewontin in the context of the IQ controversy, the question was: what is the empirical evidence for genotype-environment interaction in humans generally and IQ specifically? And during the serotonin transporter gene controversy, the question was: what is the empirical evidence for gene-environment interaction in humans generally and depression specifically? What counts as "nature" in the interaction of nature and nurture has changed from whole genomes or genotypes in the eugenics and IQ controversies to specific genes in the twenty-first century debate. And what counts as evidence of the interaction of nature and nurture has shifted from interaction in nature broadly to interaction in specific human diseases and disorders. To assess the empirical evidence in this section, I will move through each instantiation of the interaction debates and evaluate the specific formulation of the evidential question unique to that episode. The goal is to both evaluate the empirical evidence that was available at the time of the debate, and evaluate the empirical evidence that has since come to light.

The Empirical Evidence for Heredity-Environment Interaction in Nature

When Fisher and Hogben debated the interaction of nature and nurture in the early 1930s, there was very little empirical evidence to point one way or the other. Fisher undertook an investigation himself in 1923 on the relationship between different potato varieties' exposure to different fertilizers and found no interaction (Fisher and Mackenzie 1923). Hogben did not perform any experiments of his own on interaction, but he did point to the empirical results of Joseph Krafka (1920), Norman Taylor (1931), and

Frank Winton (1927), who investigated, respectively, variation in the compound eyes of flies exposed to different temperatures, variation in the sinus beat of frogs exposed to different temperatures, and variation in the mortality of rats exposed to different levels of rat poison (Hogben 1933b).

Just one decade later, a golden age of research on the interaction of nature and nurture kicked off, which persisted from the 1940s through the 1960s. The two most famous studies from this period are surprisingly unrepresentative of the broader trend, but they are still worth mentioning here. In the first, Jens Clausen, David Keck, and William Hiesey (working for the Carnegie Institute of Washington at Stanford University) investigated the height of yarrow (*Achillea*), cinquefoil (*Potentilla*), and violets (*Viola*) grown at different altitudes across California, moving from sea level into the high Sierra Nevada timberline (Clausen, Keck, and Hiesey 1940). In the second study, Roderick Cooper and John Zubek (working at the University of Manitoba) examined the spatial memory of different rats (one line bred over time to perform very well in maze tests—the "bright" rats, and one line bred over time to perform very poorly in maze tests—the "dull" rats) raised in different environments: restricted, normal, or enriched (Cooper and Zubek 1958).[3] In both cases, as can be seen in figure 6.1 below, interaction of nature and nurture was found in the populations these scientists examined. Clausen, Keck, and Hiesey found that different lines of yarrow changed their rank in terms of height quite drastically as the plants were moved from sea level, to inland, to the mountains—the Maritime variety grew taller as it was moved from a higher altitude to a lower altitude, while the Alpine variety did the exact opposite. Likewise, Cooper and Zubek found that the "bright" rats were indeed far better than the "dull" rats at the maze tests when raised in normal environments, but when these "bright" and "dull" lines were raised in the restricted and enriched environments, the gap in performance evaporated.[4]

I called the two studies above "unrepresentative" because, though widely cited as exemplifying research on the interaction of nature and nurture, they were designed to address fairly theoretical issues in biology—the

Figure 6.1

(A) Five clones of yarrow (*Achilleas*) grown at three different altitudes—Stanford (30 meters), Mather (1,400 meters), and Timberline (3,050 meters; reproduced from Clausen, Keck, and Hiesey 1940, figure 122). (B) Norms of reaction for two lines of rat ("bright" and "dull") raised in restricted, normal, and enriched environments and measured for number of errors on a maze test (y-axis; derived from Cooper and Zubek 1958).

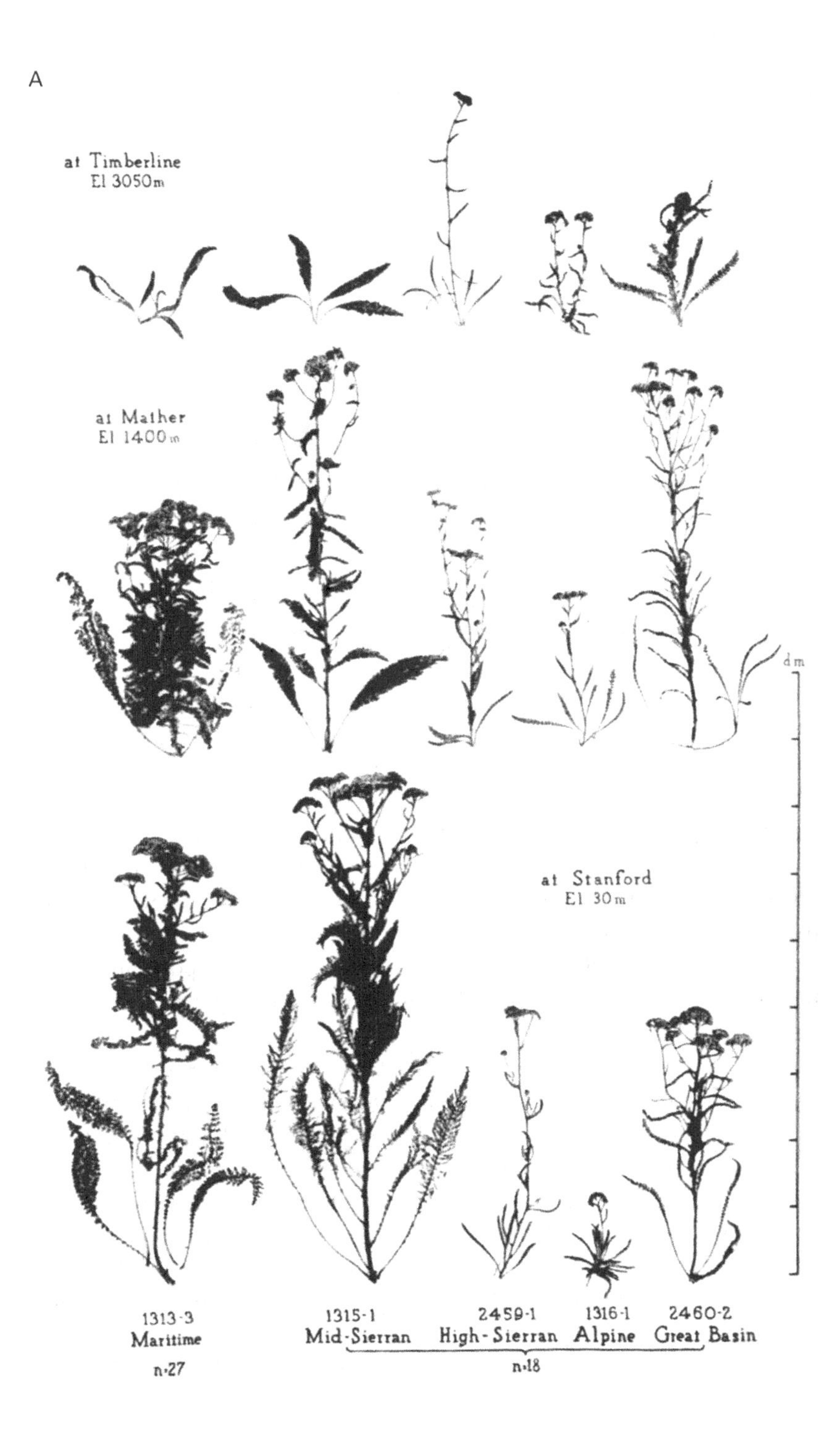
A
at Timberline
El 3050m
at Mather
El 1400m
at Stanford
El 30m
dm
1313-3
Maritime
n=27
1315-1
Mid-Sierran
2459-1
High-Sierran
1316-1
Alpine
2460-2
Great Basin
n=18

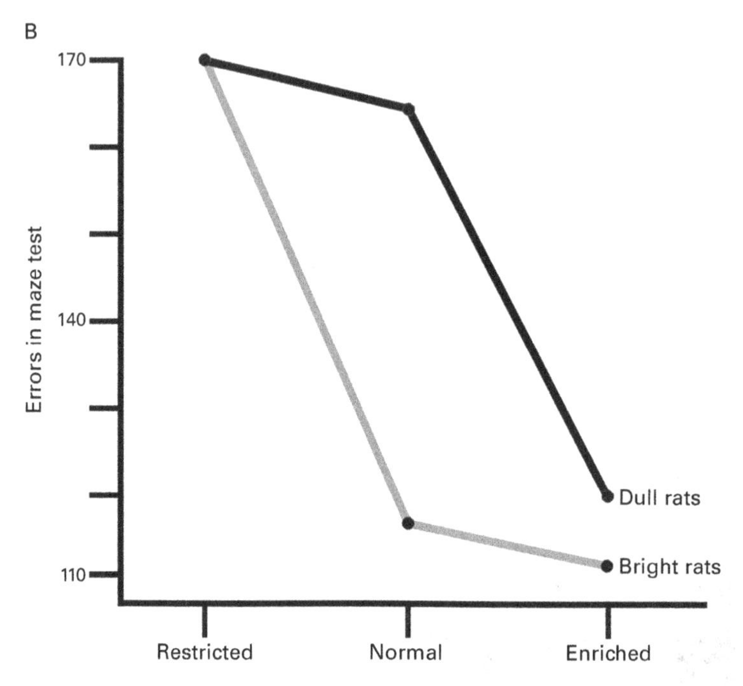

Figure 6.1
(continued)

nature of species in wild plants for Clausen, Keck, and Hiesey and the nature of learning in laboratory animals for Cooper and Zubek. In contrast, the majority of studies on the interaction of nature and nurture throughout the mid-twentieth century were of a very practical sort. The research was undertaken by agricultural scientists concerned with the possibility that the very best variety of sheep, pig, chicken, oat, barley, or corn in one environment might not be the best variety in another environment—the great complication that originally sparked Fisher's interest. For these scientists, the possibility of interaction posed a serious financial problem pertaining to the maximization of yield on investment for the farmer or agribusiness (Lerner 1950; Fitzgerald 1990). Throughout these decades, scientists reported case after case of interaction across a variety of animals and plants in journals such as the *Australian Journal of Agricultural Research*, the *Canadian Journal of Plant Science*, and the *Japanese Journal of Breeding*. In Scotland, J. W. B. King and G. B. Young reported that different varieties of sheep responded differently to different temperatures and diets when measured for wool growth (King and Young 1955). King by himself found that different varieties of pigs produced different thicknesses of streak fat (a cut of meat) when they were raised on different rations (King 1963). Pigs were also the subject of F. K. Kristjansson's study in Canada, and he found different pig varieties put on weight differently depending on whether they were fed in pastures or piggeries (Kristjansson 1957). F. H. W. Morley, in Australia, investigated assorted varieties of sheep and discovered that they responded differently to differing nutritional levels when it came to body weight (Morley 1956). And in Japan, hatch time differently affected the egg production of several chicken varieties (Yamada 1960).

Turning to plants: in the midwestern United States, J. E. Grafius described how oat and barley varieties responded differently to night temperature when it came to yield (Grafius 1956). Morley again, this time with C. I. Davern, also found differing varieties of clover flowered at different times when exposed to varying temperatures in Australia (Morley and Davern 1956). Wheat yield in Canada was differentially affected by changes in location (Baker 1969). And J. C. Murray and L. M. Verhalen found that different lines of cotton produced different levels of lint yield depending on where in Oklahoma they were grown (Murray and Verhalen 1970).

Importantly, though, the positive reports of interaction were by no means a given throughout these decades. Many of the researchers mentioned above reported cases of interaction in certain traits and, in the very same articles, noted no interaction in other traits. King, who reported differences in streak fat when it came to pigs raised on different rations, found no such interaction for shoulder, mid-back, or rump fat. Morley warned of an interaction effect on body weight in sheep exposed to different nutrition levels, but he also highlighted the fact that no interaction was found for fleece weight. Murray and Verhalen found a variety×location interaction for the lint yield of cotton grown in Oklahoma, but there was no interaction for fiber length or fiber strength in the cotton. Other researchers reported no interaction at all despite their search for it. Per Jonsson found no interaction across a range of measures for pigs exposed to varying nutritional diets in Denmark (Jonsson 1959). And King and Young, who found interaction in wool growth for different sheep exposed to varying temperatures in 1956, found no interaction when it came to fleece weight for different sheep exposed to varying nutritional diets three years later (King, Watson, and Young 1959).

When it comes to the empirical evidence for heredity-environment interaction in nature, we can sum up the situation as follows: when Fisher and Hogben were debating the evidential question in the 1930s, the question was genuinely up in the air. There were simply too few studies upon which to make any generalization. By 1969, though, the evidential landscape had changed dramatically. Empirical evidence of interaction, in part through the influential studies of Clausen, Keck, and Hiesey (1940) and Cooper and Zubek (1958), but even more so through the vast literature from agricultural scientists, was pervasive in plants and non-human animals raised in different environments. At the same time, interaction was no given. It was a genuinely empirical question, and, in a single experiment, it could be discovered for one trait and not discovered for another trait.

The best evidence for how significantly the evidential landscape changed during the mid-twentieth century comes from one-time interaction-doubters. Critics of interaction in the 1930s changed their tune by the 1950s. Fisher, for example, dismissed interaction in nature as being of "possible, but unproved importance" in 1935, and he made only passing reference to it in his *Statistical Methods for Research Workers* (Fisher 1925). But by the sixth edition of his *Design of Experiments* published in 1951, Fisher spent many pages explaining how and why to test for interaction and how to interpret the results of those tests (Fisher 1951). Likewise, Lush initially dismissed Hogben's empirical evidence of interaction as constituting only "extreme examples" (Lush 1937, 64). By the 1950s, though, Lush admitted, "Interaction between heredity and environment may be more important than most of us have thought," and he listed the interaction of nature and nurture as one of the questions most in need of attention by animal breeders (Lush 1951; Kallman 1954, see Lush's discussion at the end of Kallman 1954 on p. 172).

The Empirical Evidence for Genotype-Environment Interaction in Humans

That brings us to the IQ controversy, where the debate shifted from empirical evidence of interaction in nature to empirical evidence of interaction in humans. For plants and nonhuman animals, scientists could raise different varieties of cotton or pig in different environments in order to test for interaction, but that option obviously was unavailable for the scientist interested in studying interaction in humans. As a result, human behavioral geneticists devised a variety of methods to test for interaction in humans—identical twins reared apart, adopted children compared to their biological and adoptive parents, and identical twins compared to fraternal twins.

During the IQ controversy itself, the disputants faced the same problem encountered by Fisher and Hogben—limited empirical evidence. Lewontin, making the case for interaction, pointed to a study by Theodosius Dobzhansky and Boris Spassky (1944), but that was on flies, not humans

(Lewontin 1974). Jensen, making the case against interaction in humans, pointed to a study by John Jinks and David Fulker (1970), who surveyed the twin data available at that time (Jensen 1972). The Jinks and Fulker results, however, were not as clear-cut as Jensen made them out to be. It was true that they found no evidence of interaction for IQ (ibid., 344). They also found no evidence of interaction for neuroticism (ibid., 327). They did, however, find evidence of interaction for extraversion (ibid., 332). And they also found interaction of "borderline significance" for verbal intelligence (ibid., 336).

Jinks and Fulker's mixed bag of empirical evidence for and against genotype-environment interaction in humans was symptomatic of the research on humans throughout the 1970s, 1980s, and 1990s. For a number of traits, evidence of genotype-environment interaction was plentiful. Remi Cadoret led a series of studies in the 1980s and 1990s that investigated interaction for antisocial behavior in Iowa. In one study, Cadoret combined data from three adoption studies (Cadoret, Cain, and Crowe 1983). The adoptees were divided up based on whether or not they were exposed to genetic risk of antisocial behavior (defined by evidence of antisociality in their biological parents) and based on whether or not they were exposed to environmental risk (defined by evidence of antisociality in their adoptive home); they were then measured for antisocial behaviors in adolescence. Cadoret found evidence of interaction in all three populations of adoptees such that the presence of genetic risk or environmental risk by themselves posed little risk of antisocial behavior in adolescence, while the combination of genetic risk with environmental risk severely multiplied the risk of adolescent antisocial behavior (see also Cadoret and Cain 1981; Cadoret 1982; Cadoret et al. 1995). C. Robert Cloninger performed a similar study around this time, looking at petty criminality in a population of Swedish adoptees. And, like Cadoret, Cloninger found that exposure to both genetic and environmental risk greatly multiplied the risk of petty criminality compared to genetic risk or environmental risk alone (Cloninger et al. 1982; see also Mednick, Gabrielli, and Hutchings 1984).

Genotype-environment interaction was also reported for schizophrenia. In a Finnish population of adoptees, Karl-Erik Wahlberg and his colleagues divided up their research participants based on genetic risk and environmental risk. Genetic risk in this case meant having a biological mother diagnosed with schizophrenia. Environmental risk referred to exposure to "communication deviance" in the adoptive home; communication deviance is a pattern of communication which routinely befuddles and distracts a listener. Wahlberg discovered an extreme case of interaction in his population. Increased exposure to communication deviance severely increased the prevalence of thought-disorder (a symptomatic predictor of schizophrenia) in the adoptees already at genetic risk of schizophrenia; however, in the adoptees at low genetic risk, increased exposure to communication deviance slightly reduced the prevalence of thought-disorder (Wahlberg et al. 1997; see also Tienari et al. 1994).

Interaction was even ultimately found for IQ. For quite some time, evidence of genotype-environment interaction remained elusive. A series of studies in the 1970s and then 1980s repeatedly found no evidence of interaction for it (Plomin, DeFries, and Loehlin 1977; Plomin and DeFries 1983). The verdict on IQ changed, though, in 2003, when Eric Turkheimer (discussed in the previous section) and his colleagues found interaction involving an environmental variable directly relevant to the IQ controversy—socioeconomic status. Turkheimer's team looked at several hundred twin pairs drawn from 12 urban hospitals located across the United States, allowing them to include children from a range of socioeconomic environments. The researchers found a striking difference between the children raised in high socioeconomic homes and the children raised in low socioeconomic homes. For the children from the high socioeconomic homes, genetic variation accounted for almost all of the variation in IQ, leaving little room for environmental variation to contribute. But for the children from low socioeconomic homes, the exact opposite was the case; genetic variation contributed very little to variation in IQ, and environmental variation was the major contributor (Turkheimer et al. 2003). This result has been replicated multiple times now (Harden, Turkheimer, and Loehlin 2006; Tucker-Drob et al. 2011; for reviews of this research, see Nisbett et al. 2012 and Turkheimer and Horn 2014).

Still, as was the case with heredity-environment interaction in nature, genotype-environment interaction in humans was no given. Throughout the 1980s, Robert Plomin and his colleagues utilized the Colorado Adoption Project (a longitudinal study following several hundred adoptees for years) to investigate the evidence of interaction in humans. Plomin combined data on the children's genetic risk with their environmental exposure and looked at traits ranging from temperament and personality to behavioral problems and communicative development. The evidence for interaction was not strong. Of the literally hundreds of possible genotype-environment combinations that could have generated interaction effects, only a handful were discovered (Plomin and DeFries 1985; Plomin, DeFries, and Fulker 1988).

The evidential landscape for genotype-environment interaction in humans thus mirrors that found in plants and non-human animals. At the time of the IQ controversy itself, there was just too little empirical evidence to judge. The one study that did exist—from Jinks and Fulker (1970)—hardly settled the issue; it showed a mixed bag of interaction for some traits and lack of interaction for others. In the subsequent decades, Jinks and Fulker's results proved prescient. For some human traits, like antisocial behavior and ultimately IQ, the interaction of nature and nurture was discovered in populations ranging from Iowa to Sweden. But interaction was no given. Plomin's search for evidence of genotype-environment interaction across a range of cognitive and behavioral traits came up virtually empty.

The Empirical Evidence for Gene-Environment Interaction in Humans

In the late twentieth century, the rise of molecular biology and the human genome project allowed geneticists to switch from examining entire genomes to studying particular genes. This multiplied the pool of potential research participants enormously. Previously, if human behavioral geneticists wanted to study how nature responded to two different environments, then they needed to find identical twins reared apart. But by switching from entire genomes to particular genes, you (the reader) and I (the author), though unrelated, may share the same variant of a certain gene, and so you and I, each with our own unique environmental experiences, could be studied to see how that gene responds to different environments.

As introduced in chapter 4, this is the point in the story where Terrie Moffitt and Avshalom Caspi made headlines. They utilized the Dunedin Study, which has tracked 1,000 children born in Dunedin, New Zealand for over 40 years now. With the genetic and environmental information from these research participants combined with their medical and behavioral profiles, Moffitt and Caspi were able to investigate how different genes responded to different environments across a range of human traits. A number of their studies of gene-environment interaction followed up on the existing empirical evidence of genotype-environment interaction. In the 1980s, Cadoret and Cloninger had already found evidence of genotype-environment interaction for antisocial behavior in their populations of Iowan and Swedish adoptees. In 2002, Moffitt and Caspi identified a case of gene-environment interaction for the gene that produces monoamine oxidase A (*MAOA*), exposure to childhood maltreatment, and the subsequent development of antisocial behavior (Caspi et al. 2002; for a meta-analysis, see Taylor and Kim-Cohen 2007). In 1997, Wahlberg reported a case of genotype-environment interaction for schizophrenia in a population of Finnish adoptees. In 2005, Moffitt and Caspi published a study of gene-environment interaction for the catechol-O-methyltransferase gene (*COMT*), exposure to cannabis use in adolescence, and the subsequent development of schizophreniform psychosis (Caspi et al. 2005). Moffitt and Caspi also found cases of gene-environment interaction for depression and IQ (Caspi et al. 2003, 2007).

Importantly, Moffitt and Caspi were not the only scientists reporting cases of gene-environment interaction in humans throughout the first decade of the twenty-first century. A number of cancers were linked to instances of gene-environment interaction. Keitaro Matsuo led one study in Japan on esophageal cancer (Matsuo et al. 2001). He investigated variants in a gene—*ALDH2*—that produces aldehyde dehydrogenase-2, which metabolizes alcohol; *ALDH2* has a single nucleotide polymorphism at the 487th codon such that individuals' genomes produce either glutamine or lysine at that point in the amino acid chain. This genetic variant was combined with alcohol consumption as the environmental factor. Individuals who carried the lysine variant of *ALDH2* were at greater risk (compared to individuals carrying the glutamine variant) of developing esophageal cancer when exposed to heavy drinking, pointing to a gene-environment interaction between *ALDH2*, heavy drinking, and esophageal cancer. Cases of gene-environment interaction were also reported for bladder cancer and lung cancer (Marcus et al. 2000; Zhou et al. 2002). And, moving beyond cancers, evidence of gene-environment interaction for a range of other diseases and disorders, such as cognitive decline, spinal disc degeneration, preterm births, and eczema was also reported (Yaffe et al. 2000; Solovieva et al. 2002; Macones et al. 2004; Bisgaard et al. 2008; for additional examples, see chapter 8).

As with the empirical evidence for heredity-environment interaction in nature and genotype-environment interaction in humans, the empirical evidence for gene-environment interaction in humans was also by no means universally positive. Keitaro Matsuo (discussed just above) found a case of gene-environment interaction for the *ALDH2* gene, alcohol consumption, and esophageal cancer, but in the same study he reported no gene-environment interaction for that gene, smoking, and esophageal cancer (Matsuo et al. 2001, 914). Hans Bisgaard, in a study that combined children from Denmark and children from the United Kingdom, examined the relationship between different variants of a gene (*FLG*) that codes for filaggrin (a protein that protects the skin), exposure to cats and dogs in infancy, and subsequent development of the skin disorder eczema (Bisgaard et al. 2008). For children in both countries with a variant of *FLG* that resulted in loss of function for filaggrin, being raised in a home with cats significantly increased the risk of developing eczema within the first year of life; dog cohabitation, in contrast, had no effect on risk of eczema. Bisgaard, in short, found a case of *FLG*×cat interaction and simultaneously no *FLG*×dog interaction.

The Evidential Question: Summing Up

Recall from the chapters in part I the fairly polarized answers to the evidential question that turned up at each instantiation of the interaction debates. On the one hand, critics of interaction often went out of their way to emphasize the lack of evidence for interaction in some particular case, and then generalized from that case that interaction was rare in nature. On the other hand, advocates of interaction often went out of their way to emphasize the presence of evidence for interaction in some particular case, and then generalized from that case that interaction was common in nature.

The survey of the empirical evidence for interaction above attests to the fact that the reality is much more complex. Many studies of interaction—be it heredity-environment interaction in nature, genotype-environment interaction in humans, or gene-environment interaction in humans—have taken place over the last century. Some of these studies turned up evidence for interaction. Other studies came back negative. And quite a few of the studies turned up evidence for interaction in one trait and, in the very same population, lack of evidence for interaction in another trait. My answer to the evidential question is thus: *it's a mixed bag, and we should not assume one way or the other whether interaction exists for any particular trait or any particular gene-environment relationship*. Ask yourself: why should interaction exist for lint yield but not fiber strength in cotton? Why should interaction exist for body weight but not fleece weight in sheep? Why should interaction exist for alcohol consumption but not smoking in esophageal cancer? I see no way to justify assuming ahead of time why interaction will turn up in any given case. There is plenty of empirical evidence for it in some cases, and there is plenty of empirical evidence against it in other cases. And so when it comes to any particular case of interaction, that case simply needs to be judged on its own empirical merits.

Assessing the Serotonin Transporter Gene Controversy

That brings us to the particular case of interaction that has been by far the most controversial case in the twenty-first century—between the serotonin transporter gene (*5-HTTLPR*), stress exposure, and the development of depression.[5] As discussed in chapter 4, this case first came to prominence when Moffitt and Caspi reported that individuals with the short form of the serotonin transporter gene were at much greater risk of developing depression (compared to individuals with the long form of the gene) when exposed to repeated stressful life events such as the death of a loved one, loss of a job, or a major medical condition (Caspi et al. 2003). This original study was followed up by dozens of attempts to replicate it in populations ranging from Spain to Japan. Those replications, however, were by no means consistent; some authors reported positive replications of the original gene-environment interaction, while other authors reported failures to replicate. That mixed-bag of positive and negative replications motivated several research teams to undertake meta-analyses of the replications, compiling data from multiple studies in an attempt to assess a larger pool of data on the relationship. But rather than settling the issue, the meta-analyses only presented their own mixed bag, a not uncommon situation in biomedical research (Stegenga 2011). The meta-analyses by Kathleen Ries Merikangas and her team and by Marcus Munafò and his team came back negative (Munafò et al. 2009; Risch et al. 2009). The meta-analysis by Srijan Sen and his team came back positive (Karg et al. 2011).

In chapter 4, my goal was to explain the very different interpretations of these replications and these meta-analyses. On the one hand, advocates for research on gene-environment interaction generally and the serotonin transporter gene specifically endorsed the permissive approach to study inclusion adopted by Sen and his colleagues; they also encouraged placing the replications and meta-analyses in the context of wider biological studies that experimentally investigated the relationship between the serotonin transporter gene, stress, and depression (Moffitt and Caspi 2009; Rutter, Thapar, and Pickles 2009; Caspi et al. 2010; Kaufman et al. 2010; Rutter 2010). On the other hand, critics of research on gene-environment interaction generally and the serotonin transporter gene specifically endorsed the restrictive approach to study inclusion adopted by Merikangas and Munafò and their teams; they also dismissed as irrelevant the experimental investigations of humans, mice, and monkeys (Merikangas, Lehner, and Risch 2009; Munafò et al. 2010; Duncan and Keller 2011). My explanation of these different interpretations appealed to the explanatory divide that I tracked through each instantiation of the interaction debates in part I. If, falling on Hogben's side of the divide, you think interaction should be sought out for the important information it supplies about the causal mechanisms of the developmental process, then a focus on experimental investigations that probe the causal mechanisms and a permissive approach to study inclusion make perfect sense. Explanation, on this side, is a matter of mechanism-elucidation, wherein the challenge is to determine how genetic differences and environmental differences play out in the developmental process. If, falling on Fisher's side of the divide, you think interaction continues to bear the burden of proving itself, then a dismissal of the experimental investigations and a restrictive approach to study inclusion makes perfect sense. Explanation, on this side, is a matter of partitioning variation, wherein the challenge is to determine how much of the genetic differences and how much of the environmental differences account for the total variation in a population. On which side of the explanatory divide you fall, in short, provides different warrant for favoring the results of either of the dueling meta-analyses.

My appeal to the explanatory divide in chapter 4, while explaining the differing interpretations of the serotonin transporter gene controversy, did nothing to actually weigh in on the controversy. Recall from chapter 4 John Hardy and Nancy Low, who asked in response to the dueling meta-analyses, "What should I believe? Which explanation is true?" (Hardy and Low 2011, 455). My analysis in chapter 4 was designed to address Hardy and Low's first question; the explanatory divide points to the reason why different scientists disagreed about which meta-analysis warranted belief. I pushed Hardy and Low's second question off until this chapter. As I admitted at the end of that chapter, I intentionally held off on weighing in on the debate because I wanted to wait until I had developed my account of the explanatory bridge (in chapter 5) and my answers to the conceptual, investigative, and evidential questions (in this chapter) before turning to the debate itself. With those pieces of the puzzle now in place, let's go back to the serotonin transporter gene controversy.

Drawing on the survey of empirical evidence above, we can start by asking: should we assume depression arises from this case of gene-environment interaction, or should we assume depression does not arise from this case of interaction? The survey of the empirical evidence answers this question quite clearly: we shouldn't assume anything. Whether it was heredity-environment interaction in nature, genotype-environment interaction in humans, or gene-environment interaction in humans, the history of the empirical evidence for interaction consistently presented a mixed bag of evidence for interaction in some cases and against interaction in other cases. So to point to cases of interaction in other contexts as evidence for interaction in the serotonin transporter case or, vice versa, evidence against interaction in the serotonin transporter case as evidence against interaction in other contexts is a misguided generalization. It's not at all clear why a researcher would find interaction for streak fat but not rump fat in pigs; likewise, it's not at all clear whether the serotonin transporter gene, stress, and depression should a priori present a case of interaction.

The problem with the serotonin transporter gene controversy, of course, is that when we avoid assumptions and just turn to the empirical evidence, it presents that mixed bag of positive and negative replications and positive and negative meta-analyses. So, we are back to Hardy and Low's question: "Which explanation is true?" At present, the answer is not obvious. But that does not mean that there is nothing more to say on the issue. We can ask three related questions: what evidence should be brought to bear on the question? What is the current status of that evidence? And what might the future hold for that evidence?

Let's start by rephrasing Hardy and Low's question to get it in line with the developments of this chapter. The hypothesis under consideration is the following: the serotonin transporter gene and stress are interdependent actual difference makers in the causal mechanisms responsible for the development of depression. Call this the "*5-HTTLPR*×stress hypothesis." If the *5-HTTLPR*×stress hypothesis is correct, then evidence from variation-partitioning studies of natural human populations should pick up an interaction effect between these actual difference makers, and evidence from mechanism-elucidation studies should provide insights into how that interaction effect percolates up the difference mechanisms from the molecular, to the cellular, to the brain system, to the organismal levels. Evidence from both sets of studies is relevant and must align. You can't have an interaction effect without a difference mechanism giving rise to it. And you can't have a difference mechanism for an interaction effect that doesn't exist. So what have the variation-partitioning and mechanism-elucidation studies had to say so far about the existence of the hypothesized interaction effect and the difference mechanisms that give rise to it?

The evidence from the experimental, mechanism-elucidation studies has been largely supportive of a *5-HTTLPR*×stress hypothesis. Daniel Weinberger and his team have tracked differences in the serotonin transporter gene up through differences in amygdala structure and function (Hariri et al. 2002; Pezawas et al. 2005; see also Hariri et al. 2005; Beevers et al. 2007). Stephen Suomi, of the National Institute of Child Health and Human Development, has extended this model to non-human primates, examining how monkeys with different analogs of the serotonin transporter gene respond differently to a stressful environment (Champoux et al. 2002; see Wellman et al. 2007 for a similar study involving mice). This experimental research actively intervenes on the purported interdependent actual difference makers and tracks those interdependent actual differences up through the various levels of causal mechanism responsible for depression. So the first thing to say is that the evidence from this experimental research designed to elucidate the mechanisms is absolutely relevant to whether or not the *5-HTTLPR*×stress hypothesis is true. Looking to the future, if the hypothesis is indeed true, then mechanism-elucidation research of this sort will continue to shed new light on how differences in the serotonin transporter gene and differences in stress exposure interdependently relate to give rise to differences at the molecular, cellular, brain system, and organismal levels of depression development. In contrast, mechanism-elucidation research that finds no interaction effect would seriously question the *5-HTTLPR*×stress hypothesis. If that situation arises, a meta-analysis of just the experimental research that tests the *5-HTTLPR*×stress hypothesis would be appropriate.

The dueling meta-analyses by Merikangas, Munafò, and Sen tested whether the hypothesized interaction effect exists in natural human populations—whether there is even an effect for which the mechanism-elucidators might find a difference mechanism. The controversy, of course, follows from the fact that the first two suggest there is no effect, while the third suggests there is. But meta-analyses are only "dueling" if they truly test one and the same hypothesis. However, they did not. They did not because the Merikangas and Munafò teams restricted their meta-analyses to including only replications that (like Moffitt and Caspi's original study) combined many different stressors and then carved up the environmental variable by how many stressors were experienced by research participants. Sen's team permissively opened up their meta-analysis to allow for inclusion of replications that tested exposure or lack of exposure to a single stressor (like a heart attack). So a distinction is in order here. Merikangas and Munafo's meta-analyses tested the *5-HTTLPR*×stress_R hypothesis, where "stress_R" designates the restrictive approach they took to defining the environmental difference maker. Sen's meta-analysis tested the *5-HTTLPR*×stress_P hypothesis, where "stress_P" designates the permissive approach he took to defining the environmental difference maker.

Based on the results of Merikangas' and Munafò's meta-analyses, it is fair to say that the *5-HTTLPR*×stress_R hypothesis has been seriously challenged. But does that mean that they have simultaneously undermined the general hypothesis that the serotonin transporter gene and stress are interdependent actual difference makers? Jumping to this conclusion faces its own problem—the dozens of replications that have found the serotonin transporter gene moderates the effect of stress exposure when it comes to depression in natural human populations, and Sen's positive result that considered those studies in his meta-analysis. How do we explain all of the positive results?

One possible explanation is that all those positive replications and Sen's positive meta-analysis are simply the result of publication bias. And, indeed, critics of Moffitt and Caspi's research have said just that, arguing that journal editors are more likely to publish positive results than negative ones, and so the plethora of positive replications inaccurately represents the actual evidential landscape (Duncan and Keller 2011; Flint and Munafò 2013). The trouble with this explanation though is that, while publication bias is a general problem in science (Ioannidis 2005), Moffitt and Caspi's serotonin transporter gene study seems to be the one place where journal editors are more than happy to publish negative results, as the many published negative replications attest. So there must be something more to the relationship between the serotonin transporter gene and stress than just publication bias. The question is—what?

An alternative explanation, which takes the results from all three meta-analyses seriously, is that the general *5-HTTLPR*×stress hypothesis itself needs to be refined. These refinements can pertain to the genetic variable, the environmental variable, and the phenotypic variable, or some combination of these. The negative meta-analyses by Merikangas and Munafò followed by the positive meta-analysis by Sen speak to a potential for refining the environmental variable. Merikangas and Munafò focused on the replications that lumped many different stressful life events together, while Sen allowed for including the many more replications that focused on a single stressor. The fact that the Merikangas and Munafò teams failed to find the interaction result, while Sen's team succeeded in finding the interaction result, suggests that "stress" in the *5-HTTLPR*×stress hypothesis needs to be better defined. Consider: does it really make sense to say that someone who loses his job and then has difficulty paying his bills experienced more stress than someone who is involved in a long-term, physically abusive relationship just because "loss of job" and "difficulty paying bills" count as two stressful life events, while "involvement in a violent relationship" counts as one? With regard to the genetic variable, in 2006 a study reported that the serotonin transporter gene does not neatly divide between short and long variants because there are in fact two versions of the long variant, and one of these actually has the same reduced transcriptional efficiency as the short allele (Hu et al. 2006). As a result, the "*5-HTTLPR*" in the *5-HTTLPR*×stress hypothesis needs to take this new division into account. Finally, the phenotypic variable may need to be refined. When Moffitt and Caspi published their original result, the serotonin transporter gene and stressful life events were associated with any episode of clinical depression. George Brown, mentioned earlier in this chapter as one of the researchers who first linked stressful life events with the development of depression, recently reported a positive interaction result for the serotonin transporter gene, stress, and chronic depression (i.e., episodes lasting 12 months or more), suggesting that the *5-HTTLPR*×stress hypothesis may best apply to the most severe forms of depression (Brown et al. 2013).

To sum up, refinements to the genetic, environmental, or phenotypic variables of the *5-HTTLPR*×stress hypothesis may ultimately make that hypothesis look somewhat different from the original hypothesis reported by Moffitt and Caspi. But this should not be surprising. After all, we're not talking about the discovery of a purportedly new physical object like an asteroid, wherein scientists from diverse locations can simply point their telescopes at the same spot in the sky and confirm or disconfirm the asteroid's existence. Rather, we're talking about the potential discovery of a complicated relationship—between a gene that is still being understood at the molecular level, an environment that is difficult to quantify, and a complex psychiatric disorder. The lesson of this chapter is that, if the serotonin transporter and stress are indeed interdependent actual difference makers, then research from both variation-partitioning studies and mechanism-elucidation studies should align to support this hypothesis. Variation-partitioning studies should confirm that an interaction effect actually exists in natural human populations, and mechanism-elucidation studies should determine how the genetic and environmental variables interact during the developmental process to give rise to that interaction effect. Time will tell whether or not this alignment materializes.

III Genetic Testing for Interaction

What does the future hold for the interaction of nature and nurture? More specifically, as scientific research on the interaction of nature and nurture continues, what ethical issues will arise from attempts to draw on the results of that research to shape the world around us? The previous chapters from parts I and II considered the past and present of research on interaction. From part I we learned that, while no guarantee in any particular case, history suggests that cases of interaction will be found for a range of human traits—medical traits, cognitive traits, behavioral traits. From part II we learned that scientists will explain these human traits by accounting for variation in those traits and then elucidating the causal mechanisms of development that give rise to that variation.

One reason scientists study the interaction of nature and nurture as it relates to human traits has to do with understanding—understanding how humans come to develop the traits that they do, and how differences in those traits emerge from differences in that developmental process. Another reason has to do with intervention—to draw on that scientific understanding in order to change the world we live in by increasing the prevalence or degree of traits deemed desirable (such as cognitive abilities) and decreasing the prevalence or degree of traits deemed undesirable (such as medical diseases). In the two chapters that follow, I will investigate how research on the interaction of nature and nurture could be used to change the world we live in. These chapters are more speculative than the previous chapters, but that is by design. Where the previous chapters were designed to draw on the tools of history and philosophy of science in order to assess how we got where we are today, the subsequent chapters are designed to draw on the tools of bioethics to assess what is to come.

With cases of interaction between nature and nurture, the effects of genes and the effects of the environment are interdependent. That means that information about the status of one provides information about the potential results of the other. It also means that interventions on one influence the effect of the other. A genetic test for a gene implicated in a gene-environment interaction could guide how to intervene on the environment. Likewise, information about an environment implicated in a gene-environment interaction could suggest which gene will fare better or worse. This interdependent relationship, I hope to show, raises a unique set of ethical issues, affecting decisions ranging from which embryo to select during preimplantation genetic diagnosis to whether a dog or a cat is a better addition to a family.

7 Disarming the "Genetic Predisposition to Violence": The Dangers of Mischaracterizing Interaction

Imagine scientists discovered a *genetic predisposition to violence*. That is, imagine they identified a gene that increases the risk of engaging in violent behavior, which in turn increases the risk of engaging in criminal activity. What would you do or want done with that scientific information? For example, if you were undergoing in vitro fertilization and selecting which embryo to implant for gestation, would you want the embryos genetically screened in order to avoid implanting the embryo with the genetic predisposition to violence? Or would you want the government to screen all newborns in order to identify those with the genetic predisposition to violence for purposes of monitoring and intervention?

In 2002, a study was published that reported on the relationship between a particular gene and antisocial behaviors such as violence and criminality. The authors of the study were quickly credited with having found a "genetic predisposition to violence," and bioethical commentators warned of and even advocated for screening for the genetic predisposition. The authors of the study were Terrie Moffitt and Avshalom Caspi, whom I introduced in chapter 4 and whose study on the serotonin transporter gene, exposure to stress, and the development of depression received so much scrutiny. A year before the depression study was published, though, they published the study on antisocial behavior, and while their study of depression has received the most attention from scientists, the study of antisocial behavior has received more attention from bioethicists interested in the ethical implications of their research.

This chapter is about the results of that study on antisocial behavior and, in particular, the way in which the results of that study have been *mis*characterized, with potentially dangerous consequences. I aim to show two things: first, that Moffitt and Caspi did not identify a "genetic predisposition to violence"; discussions of their results that employ such language fundamentally mischaracterize those results. And second, this mischaracterization has in turn misdirected the bioethical discussions of the ethical implications of their research results.

In the first paragraph of this chapter, I asked what you would do or want done with scientific information about a genetic predisposition to violence. That was misleading, but intentionally so. It's fascinating to wonder what we should do with scientific information that predicts criminal violence, and that fascination is nothing new. The danger arises when the fascination distorts our understanding of the scientific results being discussed.

The Search for Biological Predictors of Criminal Violence

Efforts at identifying biological markers that are predictive of criminality and violence are not new. In the 1870s, Francis Galton in England (introduced in chapter 1 as the father of eugenics) and Cesare Lombroso in Italy both sought physically defining characteristics of the "criminal man," which would be predictive of criminal activity (Galton 1878; Lombroso 1876). Galton, for instance, developed a method he called "composite portraits." With composite portraits, a photographer takes semi-transparent headshots of many different people who all fall into the same "type" and then layers those headshots on top of one another, thus generating the composite portrait of that type. The first "type" that Galton subjected to this method was violent criminals. With the composite portrait of the violent criminal at hand, Galton surmised, "The special villainous irregularities in the [individuals] have disappeared and the common humanity that underlies them has prevailed. They represent, not the criminal, but the man who is liable to fall into crime" (Galton 1878, 97–98). Galton's reference to the disappearance of "special villainous irregularities" upon composite portraiture turned out to be more accurate than his reference to "the man who is liable to fall into crime." Despite Galton's extensive efforts, no physically defining "type" emerged for the criminally violent, as it eventually became clear that the composite criminal was essentially indistinguishable from the composite human (Pearson 1924, chapter 12).

In the 1960s, the search for biomarkers of criminal violence shifted from physical attributes to chromosomal ones. British geneticist Patricia Jacobs and her colleagues studied patients institutionalized in Scotland's high security State Hospital and reported that an unusually high number of those men carried an extra Y sex chromosome (Jacobs et al. 1965). Since females normally carry a set of XX sex chromosomes, and males normally carry the XY set of sex chromosomes, the thought was that these XYY males received an extra dose of testosterone, making them more prone to aggression than men without the extra Y chromosome. The media and the wider public were fascinated by the story; *The New York Times* ran a three-part story on the science of XYY and its implications (Lyons 1968a, 1968b, 1968c), and writer Kenneth Royce developed a popular series of novels (later turned into a television series) chronicling the tortured life of *The XYY Man* who fought to control his predisposition to criminality (Royce 1970). That fascination with a clear chromosome-crime link, however, gradually gave way to a more complicated story that suggested increased risk of criminal convictions for XYY males may be due more to hyperactivity (rather than aggression) or to lower socioeconomic resources (Rutter, Griller, and Hagell 1998, 132; Stochholm et al. 2012).

Monoamine Oxidase A and "Brunner Syndrome"

In 1978, as fascination with the XYY hypothesis was winding down, a woman approached Han Brunner at University Hospital Nijmegen in the Netherlands to seek help concerning her troublesome family (Morell 1993). Generations of males in her family, she explained, possessed a dangerous combination of low intelligence and aggressive impulsivity. One family member ran over his boss with a car; another held his sister at knifepoint and demanded that she undress; and still another stabbed the warden of a mental institution in the chest with a pitchfork. Brunner, along with his colleagues in the Netherlands and the United States, spent the next 15 years studying this family in the hopes of pinpointing the cause of the frightening behavior (Brunner et al. 1993b).

Brunner met, interviewed, and screened a number of the living males who displayed the behavior; he also had access to a family journal kept by a granduncle in the family who documented the behavior for decades among males who had by then died. The picture that emerged was of a trait that was developed by males and only certain males in the family. Women were completely unaffected, as were most of the men. But over the course of six generations, 14 males developed the combination of low IQ (roughly 85) and a proneness to aggression. By constructing a pedigree of the family and tracking the affected and unaffected men, it became clear that a recessive gene on the X chromosome was responsible for the behavior. Since women in the Dutch family (like all women) received two copies of the X chromosome, they were unaffected because they always received a dominant copy on the other X chromosome, which overrode the effects of the recessive gene. But if those same women who carried the recessive gene had children and the children were male, that meant there was then a 50% possibility that the boys would inherit the X chromosome with the recessive gene on it (Brunner et al. 1993b).

The gene in question—the *MAOA* gene—is responsible for transcribing monoamine oxidase A, or MAOA. MAOA is an enzyme in the brain that metabolizes neurotransmitters such as serotonin and dopamine, so a monoamine oxidase deficiency would lead to a chemical build-up of these neurotransmitters, which, the thought goes, leads to heightened emotional responses to stressful situations. Within the *MAOA* gene, Brunner and his colleagues identified a mutation at the 936th nucleotide. For most people, the 936th nucleotide is cytosine (C), and so it combines with an adenine (A) and a guanine (G) to make glutamine (CAG) in the MAOA molecular product. But the aggressive males in the Dutch family instead received a thymine (T) at the 936th position that in turn combined with adenine and guanine to make a "stop codon" (TAG), which instructs genetic machinery to stop producing the molecular product. The 14 males thus had a complete MAOA deficiency because of that single difference between a C and a T, and that MAOA deficiency in turn was responsible for their low intelligence and proneness to aggression (Brunner et al. 1993a).

Brunner was hailed as having found the "aggression gene" and the "warrior gene" (Morell 1993; Gibbons 2004). But this was an honor that he actively fought, and for good reason (Brunner 1996). Brunner and his colleagues identified an extremely rare mutation (C to T). This rare mutation led to an extremely rare condition—complete MAOA deficiency, which came to be called "Brunner syndrome." And this rare condition was isolated in a single Dutch family. To this day, only 14 men are documented to have been afflicted with Brunner syndrome—the 14 men first brought to Brunner's attention back in 1978. So to the extent that aggression and criminal violence are a general problem for society, Brunner syndrome certainly provides no explanation for it.

From Brunner Syndrome to the Interaction of Nature and Nurture

Although Brunner syndrome would not explain aggression and criminal violence in general, it did suggest looking at MAOA and the *MAOA* gene. Terrie Moffitt and Avshalom Caspi drew on the Dunedin Multidisciplinary Health and Development Study (or just "the Dunedin Study") to examine that possibility. Researchers with the Dunedin Study, as introduced in chapter 4, have been tracking roughly 1,000 children born in Dunedin, New Zealand, regularly since their births in 1972–1973. Over the course of four decades, the researchers have gathered a wealth of information about the study participants' environments, genes, and traits. Moffitt and Caspi,

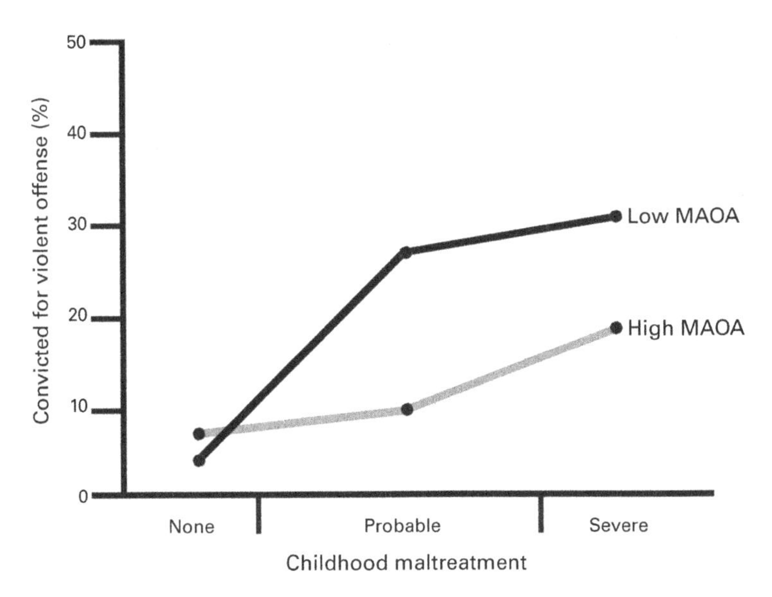

Figure 7.1
Norm of reaction graph for *MAOA*, childhood maltreatment, and violent offense convictions (adapted from Caspi et al. 2002, figure 2B).

in an effort to better understand criminal violence, combined *MAOA* with a known environmental cause of antisocial behavior—exposure to childhood maltreatment, such as physical abuse, sexual abuse, and neglect (Caspi et al. 2002). And rather than looking for a rare genetic mutation (Brunner's focus), Moffitt and Caspi instead focused on natural variation in the *MAOA* gene. The *MAOA* gene comes in relatively short and long forms, and the result is relatively low and high MAOA activity. They combined this difference between low MAOA and high MAOA with different degrees of exposure to childhood maltreatment: none; probable (defined as evidence of exposure to one example of childhood maltreatment); and severe (defined as evidence of exposure to two or more examples of childhood maltreatment). As can be seen in figure 7.1, Moffitt and Caspi found that males from the Dunedin Study who carried the low-MAOA variant of the *MAOA* gene were much more vulnerable to the effects of childhood maltreatment. Low-MAOA males who were maltreated were more likely to be convicted of violent offenses than high-MAOA males who were maltreated (figure 7.1). Moreover, on a composite measure of antisocial behavior (which combined information from criminal records, a psychiatric evaluation, and reports from people who know the Dunedin participants), low-MAOA males who were maltreated were much more likely to exhibit signs of antisocial behavior than high-MAOA males who were maltreated (see figure 1 in Caspi et al. 2002).

Moffitt and Caspi's *MAOA* study, in contrast to Brunner's *MAOA* study, potentially offered a much more generalizable insight into the causes of aggression and criminal violence in society. In contrast to the rare genetic mutation in the Dutch family that Brunner studied, Moffitt and Caspi investigated naturally occurring variation in the *MAOA* gene. So while I hope I can safely assume that I don't suffer from Brunner syndrome, I could be genetically screened for my *MAOA* status, and I would be either low MAOA or high MAOA depending on which variant of the gene that I carry. (Approximately one-third of males are low MAOA and approximately two-thirds are high MAOA; Foley et al. 2004; Kim-Cohen et al. 2006.) What's more, by combining the genetic information about *MAOA* with the environmental information about exposure to childhood maltreatment, Moffitt and Caspi were able to account for much more natural variation in antisocial behavior than if they had looked at either the genetic or the environmental factor in isolation. Indeed, the combination of *MAOA* with childhood maltreatment produced some truly frightening results. In figure 7.1, for instance, only 12% of the Dunedin participants fell into the category of low-MAOA/severe-maltreatment, but that 12% accounted for a full 44% of the male population's violent convictions. Moreover, 85% of the participants who carried both the low-MAOA variant and were severely maltreated displayed some form of antisocial behavior. In short, individuals with the low-MAOA variant and who experienced severe maltreatment presented a genuinely troubling population—they stood a better than 8 in 10 chance of showing some sign of antisocial behavior, and even though they made up a very small portion of the overall population, they accounted for almost half of that population's violent convictions (Caspi et al. 2002, 853). These numbers fascinated bioethical commentators.[1]

Intervening on the "Genetic Predisposition to Violence"

Almost immediately after Moffitt and Caspi published their study, bioethicists from philosophy, medicine, and law began considering the ethical implications of their discovery. The thread that united these bioethical commentaries was the claim that Moffitt and Caspi found individuals who carried the low-MAOA variant to be "genetically predisposed to violence." The question then became: what should be done with information about this genetic predisposition? Bioethicists identified two venues where something could be done: (a) parents interested in preimplantation genetic diagnosis during in vitro fertilization with an eye toward avoiding implantation of the low-MAOA embryos, and (b) states interested in a policy of newborn screening with an eye toward monitoring and intervening on the low-MAOA children.

Preimplantation Genetic Diagnosis: Screening against Low-MAOA Individuals

During in vitro fertilization, it's possible to genetically screen embryos to decide which embryo or embryos to implant for gestation. For families with a history of Huntington's disease, for instance, preimplantation diagnosis offers prospective parents the opportunity to ensure that they do not have a child with the debilitating neurodegenerative condition (Sermon et al. 1998, 2002). A number of ethical commentators, though not endorsing the preimplantation genetic diagnosis of embryos for their *MAOA* status, simply flagged the issue as one bioethicists should watch and consider. For example, Jonathan Moreno warned, "Prospective parents might therefore test embryos for the MAOA marker before implantation to avoid giving birth to a child with this particular potential for criminality" (Moreno 2003, 151). Likewise, David Wasserman titled an article "Is There Value in Identifying Individual Genetic Predispositions to Violence?" (Wasserman 2004). Since Moffitt and Caspi found that the 12% of their Dunedin Study participants who were both maltreated and carried the low-MAOA variant accounted for 44% of the male population's violent offense convictions, Wasserman wondered, "[W]hat if we could identify some individuals in that 12% not only at birth, but in utero, or before implantation?" (ibid., 24).

Julian Savulescu, in contrast to Moreno and Wasserman, has gone further. Rather than just flag the preimplantation genetic diagnosis of *MAOA* as a potential bioethical issue, Savulescu has advocated for such genetic screening as "morally required" of parents. The moral requirement follows from his "principle of procreative beneficence." Savulescu's principle of procreative beneficence offers the following guideline: "couples (or single reproducers) should select the child, of the possible children they could have, who is expected to have the best life, or at least as good a life as the others, based on the relevant, available information" (Savulescu 2001a, 415). When Savulescu first introduced his principle of procreative beneficence in 2001, Moffitt and Caspi had not yet published their 2002 study on antisocial behavior, and so Savulescu did not have the results from their study at hand. However, even at that earlier date, Savulescu had his eye on a genetic predisposition to violence as a perfect case wherein his principle of procreative beneficence would guide parents away from selecting an embryo that would be plagued by a life of aggression. "Imagine there is a gene which contributes significantly to a violent, explosive, uncontrollable temper, and that that causes people significant suffering," he wrote, and then assessed: "Violent outbursts lead a person to come in conflict with the law and fall out of important social relations" (ibid., 420). The resulting loss of independence, dignity, and important social relations, Savulescu argued, would prohibit the carrier of the genetic predisposition from achieving the best life relative to those who did not carry the predisposition, and so parents should select the predisposition-free embryo (ibid.; see also Savulescu 2001b).

When Savulescu returned to the issue in 2006, Moffitt and Caspi's results were then available, and they figured prominently for him alongside Brunner's study of the Dutch family. The two studies—Brunner's on the Dutch family and Moffitt and Caspi's on the Dunedin population—together supported a "monoamine oxidase hypothesis": "In such cases, the genetic mutation, though perhaps not a cause, can be seen as a significant risk factor for some kinds of violent criminal behaviour" (Savulescu et al. 2006, 160). So rather than advocating for selecting against any old genetic predisposition to violence, Savulescu now identified the "MAOA mutation" as the target of preimplantation genetic diagnosis.

Newborn Screening: Monitoring and Intervening on Low-MAOA Individuals

States commonly mandate the screening of all newborns for a set of diseases and disorders (the set varies from state to state). For example, newborn screening for phenylketonuria (or "PKU") identifies newborns who, for genetic reasons, cannot metabolize phenylalanine—an amino acid commonly found in food; by identifying these cases at birth, the children can be put on a phenylalanine-free diet and thus avoid developing the debilitating symptoms of PKU. Newborn screening is another venue where bioethicists have drawn attention to the implications of Moffitt and Caspi's study. And, as with preimplantation genetic diagnosis, several commentators have simply pointed to the possibility of newborn screening for *MAOA* as an issue of bioethical significance. Ravinesh Kumar, for example, suggested that "although likely to be controversial, we may decide to identify 'at-risk' individuals prone to developing antisocial tendencies by screening for the MAOA functional polymorphism and recognizing those who harbor the 'low-activity' variant" (Kumar 2003, 184). Paul Appelbaum similarly warned that "the pressure to screen is likely to increase if intervention can be shown to actually reduce crime. If effective treatment becomes available, the pressure to identify [at-risk] individuals through screening at birth may be irresistible" (quoted in Moran 2006). And Wasserman again noted,

> It would be tempting for the state to set a low threshold for pharmacological prevention for large categories of potentially vulnerable young men. For example "prescribing" MAOA supplements for all young men screened as low MAOA, regardless of social class or family circumstances, would avoid blatant class bias, as well as intrusive and expensive inquiries into maltreatment [that] fell short of abuse and neglect. (Wasserman 2004, 28)

Jessica Brooks-Crozier, in contrast to Kumar, Appelbaum, and Wasserman, has gone further. For, like Savulescu, Brooks-Crozier did not just flag the issue; instead she advocated for it. Brooks-Crozier has made a case for the mandatory screening of newborns to identify those with low MAOA (Brooks-Crozier 2011). On her proposal, all children would be genetically screened for their *MAOA* status within 48 hours of birth, and parents would not have the option to opt out of this test. In the case of an "abnormal test result" (i.e., low MAOA), the child's healthcare provider and the state's department of health would be notified. An individualized family service plan would then be developed for the low-MAOA child and family designed to prevent the development of antisocial behavior; this plan would include support services such as home visits from social workers to provide education, counseling, and monitoring (ibid., 542–546). Importantly, Brooks-Crozier did not offer this proposal as some far-fetched vision for the distant future; rather, she argued that there were strong legal grounds for implementing the proposal right now:

> The "disorder" occurs frequently in the population, can be screened for, and is treatable. The threat posed by MAOA-low males is real and substantial and the means adopted by states to confront that threat—mandatory newborn screening for the MAOA-low genotype followed by intervention services—is likely to be effective. Thus a court is likely to find that the state interest in mandatory newborn screening for the MAOA-low genotype is compelling. (ibid., 551)

From a "Genetic Predisposition" to a Differential Susceptibility

I'm now going to argue that the overarching conceptual framework the authors above have used to diagnose the bioethical issues is misguided. It is misguided because the authors' bioethical analyses follow from first assuming that Moffitt and Caspi identified a "genetic predisposition to violence." They did not.

Interaction as a Change in Scale

To begin seeing why Moffitt and Caspi did not identify a genetic predisposition to violence, it is necessary to see how cases of gene-environment interaction can come in two forms: those that result in a *change in scale*, and those that result in a *change in rank*. In general, the interaction of nature and nurture refers to cases where different genetic groups (like low MAOA and high MAOA) respond differently to an array of environments (like various degrees of childhood maltreatment).[2] But the key is that "respond differently" comes in degrees. The differential response can be mild or extreme. On the mild end, a case of interaction that results in a change of scale refers to a situation where the different genetic groups respond differently to an array of environments, but that differential response does not alter the fact that the higher-ranking genetic group maintains its higher ranking across all of the environments (Lynch and Walsh 1997). A norm of reaction graph will help to visually convey this idea, so consider the one in figure 7.2. In figure 7.2, I took Moffitt and Caspi's original norm of reaction graph for *MAOA*, childhood maltreatment,

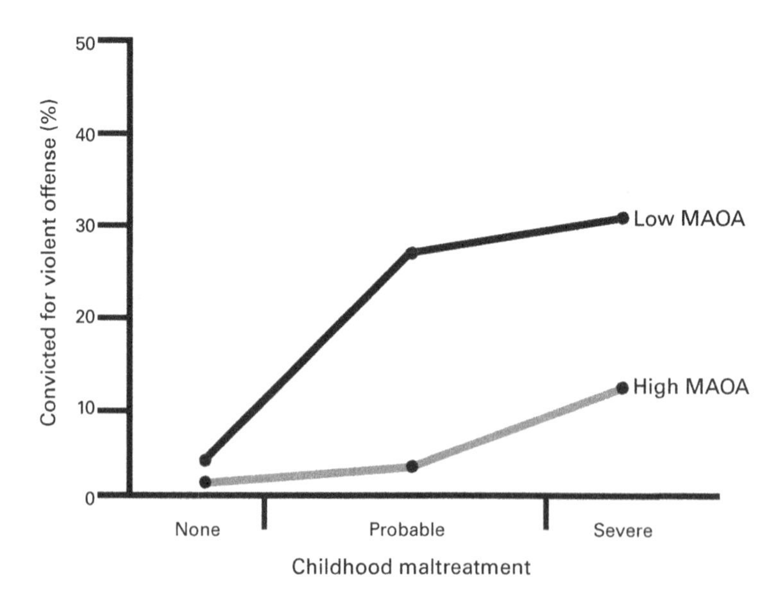

Figure 7.2
A hypothetical norm of reaction graph for *MAOA*, childhood maltreatment, and violent offense convictions.

and violent offense convictions (figure 7.1 above), but I altered it in an important way: I reduced the measure of convictions for the high-MAOA group in each environment by about 7%. So now the low-MAOA group maintains its higher ranking on the measure of violent crime convictions across all three environments. This is still a case of gene-environment interaction because the low-MAOA and the high-MAOA groups do still respond differently to the array of environments (the high MAOA increases gradually, while the low MAOA increases drastically), but all that has changed is the scale of the difference between the two groups in the different environments; the difference gets bigger as childhood maltreatment is encountered.

To evaluate the appropriateness of the concept of a genetic predisposition as it applies to cases of interaction, a definition of this concept must first be afforded. A genetic predisposition can be defined as follows:

> The presence of a genetic difference between various groups *consistently* increases the probability of individuals from one group, in comparison to individuals from the other group(s), developing a particular trait *regardless of* the measured environmental condition.

Note the relational nature of this definition. Members of any genetic group may be susceptible to developing the particular trait under investigation if exposed to the environmental condition. But attaching "genetic" to "predisposition" is only appropriate if a *genetic* difference consistently increases the probability of individuals from one group developing the trait relative to individuals from the other group(s).

Cases of gene-environment interaction resulting in a change of scale may be appropriately characterized with the concept of a genetic predisposition as defined above. Consider the hypothetical case graphed in figure 7.2: in every measured environment (none, probable, and severe childhood maltreatment), individuals in the low-MAOA genetic group maintained their relatively elevated risk for violent offense conviction. What "genetic predisposition" implied in this case, then, was that the presence of the genetic difference between the two groups consistently put individuals from the low-MAOA group at an increased risk of being convicted of a violent offense relative to the individuals from the high-MAOA group. Or, for an empirical example, consider the *BRCA1* and *BRCA2* genes responsible for increased risk of breast and ovarian cancers. The phrase "a genetic predisposition to breast and ovarian cancer" is appropriately linked with these genes because bearing these genes consistently increases the risk of developing breast and ovarian cancer in the known environments.

Interaction as a Change in Rank

But notice that the above account is decidedly not what occurs in the actual results from Moffitt and Caspi's study. Their *MAOA* study is instead an instance of interaction resulting in a change of rank. An instance of gene-environment interaction resulting in a change of rank refers to a situation in which different genetic groups respond differently to the same array of environments, and that differential response is so extreme that the higher-ranking group in one environment becomes the lower-ranking group in a different environment—where the norms of reaction actually *cross over* (Lynch and Walsh 1997). Notice that this crossover effect is precisely what we find in the actual data from Moffitt and Caspi's low- and high-MAOA groups graphed above in figure 7.1. In the environments with probable and severe childhood maltreatment, the low-MAOA group did in fact have higher percentages of violent offense convictions and did in fact score higher on the composite index of antisocial behavior than the high-MAOA group. But in the environment with no childhood maltreatment the exact opposite was the case—the low-MAOA group had a lower percentage of violent convictions and scored lower on the antisocial behavior index.

The concept of a genetic predisposition captures cases where a genetic difference between groups consistently increases the probability of individuals from one of these groups developing a trait regardless of the measured environment. But in Moffitt and Caspi's study, the environmental conditions were crucial for assessing the relationship between the low-MAOA and the high-MAOA groups with regard to risk of violent crime convictions and antisocial behavior. Prior to an individual actually experiencing (or otherwise predicting) exposure to childhood maltreatment, there is simply no way to assess whether an individual with low-MAOA activity will be more or less prone to antisocial behavior than an individual with high-MAOA activity. The low-MAOA individual is less likely to be convicted of a violent offense in environments with no childhood maltreatment and simultaneously more likely to be convicted of a violent offense in environments with probable and severe childhood maltreatment. Employing the concept of a "genetic predisposition to violence" when the environmental conditions are unknown, we are forced incoherently to say that individuals in the low-MAOA group are simultaneously more prone to violent crime and, at the same time, less prone to violent crime. In short, the concept of a genetic predisposition fundamentally mischaracterizes these cases of interaction that result in a change of rank precisely because it leads to this incoherent result.[3]

What we need, then, is a new concept—a concept that is specifically tailored for these cases of interaction that result in a change of rank, and a concept that sets these cases (and the bioethical implications of these cases) apart from cases of genetic predisposition or genetic determinism. Elsewhere, I suggested using "interactive predisposition" to characterize the phenomenon (Tabery 2009c, 2009d). However, after introducing the concept of interactive predisposition, it came to my attention that developmental psychologist Jay Belsky had already offered a new term to capture such cases of interaction that change in rank—"differential susceptibility." Belsky's focus has been on the evolutionary implications of gene-environment interaction. Interactions that change in rank are exactly what we should expect in nature, Belsky has counseled; in a world where it is not always easy to predict what the future holds, differential susceptibility is evolutionarily advantageous because it allows a population to respond to a range of environmental exposures rather than go "all in" on one environment (Belsky 1997; Belsky, Bakermans-Kranenburg, and van IJzendoorn 2007; Belsky et al. 2009; Belsky and Pluess 2009). Science writer David Dobbs, who has been tracking research on the differential susceptibility idea for quite some time, summed up the idea nicely in a 2009 piece for *The Atlantic*: it creates a diversified-portfolio approach to nature (Dobbs 2009).[4]

Despite our distinct interests (mine on the bioethical implications and Belsky's on the evolutionary implications), the concepts of interactive predisposition and differential susceptibility are essentially synonymous. So to respect Belsky's priority, I will proceed in the pages that follow to use his term. A differential susceptibility may be defined as follows:

> The presence of a genetic difference between various groups both *increases* and *decreases* the probability of individuals from one group, in comparison to individuals from the other group(s), developing a particular trait *depending on* the measured environmental condition.

Where a genetic predisposition is relational in one sense, a differential susceptibility is relational in two senses. Like the concept of a genetic predisposition, the concept of a differential susceptibility is relational in the sense that the probability of individuals from one group developing the trait under investigation is always considered in comparison to individuals from the other group(s) developing the trait. For a genetic predisposition, however, that relation between the groups maintains a consistency (between which is higher and which is lower ranking) across all measured environments, whereas this is not the case for a differential susceptibility. For a differential susceptibility, the relation between the groups is itself relative to which environment they're in.

A Question (or Qualification) of Significance

In contrast to Belsky's focus on the evolutionary implications of interactions that result in a change of rank, I want next to focus on the bioethical implications. Before proceeding to that discussion, though, a qualification is in order. Interestingly, while the norms of reaction in figure 7.1 do clearly cross over each other (or change in rank), the difference between the low-MAOA and the high-MAOA groups in the environment without childhood maltreatment is not statistically significant. Recall that R. A. Fisher first introduced tests of statistical significance when he studied the response of different potato varieties to different fertilizers (Fisher and Mackenzie 1923). A statistical significance test judges whether an effect found in a study is likely to reflect what's actually going on out there in the world or whether the effect is just the result of chance. If the result is statistically significant, then that means the effect is unlikely to be just a chance finding; if the result is not statistically significant, then that means the effect could just be due to chance. So when I said above that the difference between the low-MAOA and the high-MAOA groups in the environment without childhood maltreatment was not statistically significant, that meant the difference between those groups could be just a chance finding. If it was a chance finding and there was no real difference between the two genetic groups, then my analysis in this chapter would be inapplicable to the case of *MAOA*, childhood maltreatment, and antisocial behavior. It would be inapplicable because the case would essentially be reduced to a case of interaction that results in a change of scale (not rank), and so it would be legitimately described as evincing a genetic predisposition (not a differential susceptibility).

However, while I admit that the qualification above is in order, I simultaneously argue that it is unlikely that the difference between the low-MAOA and the high-MAOA groups in the environment without childhood maltreatment is simply the result of chance. The reason is that, in the replications of Moffitt and Caspi's original study and in meta-analyses of those replications, the change in rank keeps coming up; it's robust (see, for instance, Foley et al. 2004; Haberstick et al. 2005; Nilsson et al. 2006; Kim-Cohen et al. 2006). A better explanation for the lack of statistical significance, thus, is insufficient sample size. It is notoriously difficult to detect interaction effects without very large sample sizes (Wahlsten 1990). Moffitt and Caspi's original result, as well as the subsequent replications, have data from research participants spread out across all of the gene-environment combinations, which means there is just a fraction of that larger set in any particular gene-environment combination, which means that looking for a statistically significant difference between those subcombinations becomes difficult indeed. As a result, the repeated findings of a crossover effect suggest that the change in rank is robust, and so I will proceed to the bioethical discussion as if this is the case.

Intervening on the Differential Susceptibility

I hope to have now conveyed why it is inaccurate to say that individuals with low-MAOA have a "genetic predisposition to violence." It is inaccurate because, while Moffitt and Caspi's study does suggest that low-MAOA individuals are at a greater risk of antisocial behavior and violent crime if they are maltreated during childhood, the study also suggests that those same low-MAOA individuals are at less risk when no childhood maltreatment is experienced. With this basic point in place, let's now return to the two venues where bioethicists have suggested something could be done with genetic information about *MAOA*: preimplantation genetic diagnosis and newborn screening.

Preimplantation Genetic Diagnosis: The Danger of Screening against Low-MAOA Individuals

Might parents want to screen their embryos during in vitro fertilization for *MAOA* status in order to avoid implanting an embryo with low MAOA? Jonathan Moreno and David Wasserman warned that they might. Julian Savulescu argued that they should. Well, parents should do this *only if they are already planning on maltreating their children!* That is, if the children are not likely to experience childhood maltreatment, then it is in fact the low-MAOA embryos that are less likely than the high-MAOA embryos to develop into adults with antisocial behavior. This confusion arises from assuming that you can predict which genetic group is at greater risk without any information about what environment will be experienced. But, with cases of interaction that change in rank, that is simply a myth. With cases of interaction that change in rank, risk is dependent upon both the genetic difference and the environmental difference, and so any predictive effort must include information about both.

Newborn Screening: The Danger of Monitoring and Intervening on Low-MAOA Individuals

Might states want to screen all newborns for their *MAOA* status in order to monitor and intervene on those children with low-MAOA? Kumar, Appelbaum, and Wasserman warned that they might. Brooks-Crozier argued that they should. Two possibilities for intervention have been envisioned: pharmaceutical and social-behavioral. In both cases, however, the change in rank between the low-MAOA and the high-MAOA individuals means intervening on the low-MAOA children who are never maltreated won't decrease their risk; it will increase it.

Take the pharmaceutical intervention. The thought here seems to be something like an "MAOA booster shot." Since the low-MAOA children who are maltreated subsequently develop an increased risk of antisocial behavior, the idea would be to supply them with extra MAOA from birth in order to pharmaceutically switch them to high-MAOA status. By switching to high-MAOA, the thought goes, they would now be protected against the antisocial possibility associated with childhood maltreatment. The problem with this proposal, though, is that it involves giving the MAOA booster shot to all newborns who are low MAOA. And a glance back at figure 7.1 will show that, while this MAOA boost will reduce the risk of antisocial behavior and criminal violence in the children who go on to be maltreated, it will simultaneously increase the risk of antisocial behavior and criminal violence in the children who are never maltreated. What's more, in the Dunedin population two-thirds of the males fell into an environment with no maltreatment, while the other one-third were spread out over the environments with probable and severe maltreatment. This makes sense; most children are not maltreated. What this proposal would mean, then, is that two-thirds of the population would have their risk of antisocial behavior slightly increased in exchange for the other one-third having their risk significantly decreased. With more people standing to undergo harm from such a pharmaceutical intervention than standing to benefit, the burden is on those envisioning such an intervention to make the case for its implementation.[5]

What about Brooks-Crozier's suggested social-behavioral intervention? On her proposal, recall, all children would be screened for their *MAOA* status at birth, and an "abnormal test result" (i.e., low MAOA) would be reported to the child's healthcare provider as well as the state's department of health; then an individualized family service plan would be developed for the low-MAOA child and family designed to prevent the development of antisocial behavior. While I applaud Brooks-Crozier's appeal for increased social services aimed at preventing the abuse and neglect of children, it is the particular way in which she targets all and only low-MAOA children that worries me. The social-behavioral services she envisions are not offered to all families; they are only offered to the families with children who have the "abnormal test result." This selective intervention raises the very real possibility that children with the "abnormal test result" will come to think they are, well, "abnormal." The parents are also likely to internalize this worry; after all, their child screened positive for an "abnormal test result," and now social services are visiting regularly to prevent criminal violence from materializing in their child. The worry here is of a self-fulfilling prophecy—of a child and family expecting and thus facilitating the development of an antisocial personality by virtue of the very diagnosis and intervention designed to prevent it (Davison 1996; Campbell and Ross 2004).[6]

Let's now think about how the cost of a self-fulfilling prophecy would weigh up against the benefit of an intervention program designed to prevent the development of antisocial behavior. Imagine for a moment that the leaders of Dunedin enacted Brooks-Crozier's proposal and began screening all of their newborns for *MAOA* status; suppose also that the percentages of maltreatment Moffitt and Caspi tracked for the original Dunedin participants continued to hold. One-third of the low-MAOA children would have been maltreated had the proposal not been implemented, but if the intervention is successful then those children stand to benefit quite a bit—now they are not maltreated, and now they are at less risk of antisocial behavior and violence. For these children we might say that the potential danger of a self-fulfilling prophecy is well out-weighed by the upside of not developing antisocial behavior. But what about the other two-thirds of the low-MAOA children? They weren't going to be maltreated in the first place. As a result, they bear all the burden of the self-fulfilling prophecy without any of the upside of the intervention. So, as with the pharmaceutical intervention, we are back in a situation where the proposed intervention stands to harm more of the children subjected to the intervention than it helps. This is the challenge posed by intervening on the genetic side of the differential susceptibility.[7]

An Alternative: What about Intervening on the Environmental Side?

All the attention to a genetic link between *MAOA* and antisocial behavior has distracted from an alternative target for intervention—the environment. Childhood maltreatment, be it severe forms like physical and sexual abuse or less severe forms like neglect and harsh punishment, are damaging to all children and dangerous to the society that receives these children after they leave home (U.S. Department of Health and Human Services 2012). Brooks-Crozier suggested offering a social-behavioral intervention for all and only children born with low-MAOA. But what about, rather than intervening on the genetic factor, intervening on the environmental one by extending that service to all children and their families? Brooks-Crozier considered the cost associated with applying her program to children born in Connecticut in 2009; with 40,000 children born, her program would be extended to 6,600 low-MAOA children and families at a cost of just over $87 million. In Connecticut in 2009, however, there were also 5,760 aggravated assaults, 651 rapes, and 107 murders, which she estimated cost the state almost $3 billion. Brooks-Crozier's proposal surely would not prevent every violent crime, as she readily admitted; crimes were also perpetrated by high-MAOA individuals, and the intervention would not be 100% successful. But her argument was that it could put a serious dent in the crime rate, and that dent could save the state a huge amount of money. So then: why only the low-MAOA children? What about simply extending the social-behavioral intervention to all 40,000 children? The financial cost would obviously go up: to over $1 billion. But the reach of the intervention would also go up, and it would still be well under the $3 billion associated with violent crime in the state. In exchange, there would be no financial cost associated with an additional genetic test and no social-psychological cost associated with a self-fulfilling prophecy because all children and families would receive the intervention.

Now, to be clear, I am not advocating for this particular environmental intervention. There are obvious financial concerns about paying for such a major initiative as well as political concerns about getting the government involved in something as personal as parental decision making. I'm simply pointing out that intervening on the genetic variable in these cases of interaction that change in rank is particularly difficult because the intervention stands to decrease risk for some while increasing risk for others. But intervening on the environmental variable presents no such difficulty in this case. Childhood maltreatment, after all, does not increase risk of antisocial behavior for some and decrease it for others; it presents a genuine environmental predisposition for all, regardless of *MAOA* status (Kim-Cohen et al. 2006).

The irony is that, while the *MAOA*×maltreatment case of gene-environment interaction has received the most attention from bioethicists, it's actually a rather simple case when it comes to making the case for an intervention. Everyone can agree that childhood maltreatment is bad, and childhood maltreatment is bad for everyone (regardless of whether or not it leads to violence and antisocial behavior). So intervene on the environment and intervene for all. There are other case of interaction that change in rank, however, where this isn't so clear-cut—cases where the environmental variable is less obviously harmful and cases where the environmental variable does both increase and decrease risk. It is to these trickier cases that I turn in chapter 8.

8 Of Dogs, Daycare, and Discipline: A "Genetic Guide to Parenting"?

Consider the following three questions:

- Should parents buy their child a dog or a cat?
- Should parents send their infant to daycare?
- Should parents, when their toddler has a temper tantrum, respond with empathy directed at the child's frustration or with punishment directed at the child's misbehavior?

Answering these questions is complex. In contrast to childhood maltreatment (discussed in the last chapter), which everyone can agree is harmful, decisions about the family pet, childcare, and discipline are subjects of friendly debate and involve both personal and familial evaluations of finances, schedules, and parenting philosophies.

Might genetic information about the child figure into such decisions?

I believe it could, and in the near future. My goal in this chapter is to say why I think a "genetic guide to parenting" is a very real possibility and to then evaluate whether or not this would be a good thing. Incorporating genetic information into traditional parental decision making, I hope to show, will fundamentally alter what counts as a "treatment." This alteration would have both advantages and disadvantages. On the one hand, it would empower parents by putting treatment, which was previously the exclusive purview of medical professionals, directly into the hands of parents. On the other hand, it would transform the way parents conceive of certain parental decisions, reshaping how parents think about the rationale for various childcare choices.

Differential Susceptibilities Involving the Early Childhood Environment

My reference just above to a genetic guide to parenting might cause the reader to think I've suddenly forgotten the lessons of the last seven chapters about the interaction of nature and nurture and slipped into a perspective of genetic determinism. But that is not the case, for the scientific research I will discuss below involves cases of gene-environment interaction, wherein interaction effects have been reported between specific genes and specific environments. What makes these cases interesting and might make for a genetic guide to parenting comes from the fact that the environmental variables in the cases are actively shaped by parents when they make decisions about the exposures and experiences of their child's early life. All of the cases involve interaction effects that are so extreme that the riskier genetic group in one environment becomes the least risky group in a different environment; the cases, that is, all involve interactions that change in rank, constituting differential susceptibilities to the traits in question.[1]

Allergies: Dogs, Daycare, and the *CD14* Gene

"Atopy" derives from the Greek word for being out of place. People with atopic syndrome are hypersensitive to common environmental antigens. This hypersensitivity leads to an overactive immune response, resulting in a range of allergic symptoms: atopic dermatitis (more commonly called "eczema"), allergic rhinitis (more commonly called "hay fever"), allergic conjunctivitis (more commonly called "red eyes"), allergic asthma, and food allergies.

Hundreds of millions of people worldwide suffer from allergies. According to the United States' National Institute of Allergy and Infectious Disease, 7 million children in the United States are asthmatic. 30% of the U.S. population (mostly children and adolescents) suffers from eczema. Respiratory allergies, such as hay fever, affect 10% of U.S. children. And 5% of children in the United States have food allergies. Many of these children with allergies will grow into adults with allergies. And as anyone with allergies can tell you, allergic symptoms can be seriously debilitating. Quality of life is diminished when people with allergies avoid certain environments that might trigger an allergic response. Serious episodes, such as asthma attacks, can even result in death. The financial toll is significant—every year, billions of dollars are either spent on medical care for allergies or lost due to missed work and school days.[2]

The story of allergy genetics over the last two decades is quite similar to the story of behavioral genetics over that same period. As told in chapter 4, initial discoveries of candidate genes associated with behavioral traits in the 1990s brought great excitement, but that excitement gave way to disappointment when many of the original candidate genes failed to replicate in new populations. In chapter 4, remember, Moffitt and Caspi sought to move past the direct gene-behavior link by incorporating environmental variables such as breastfeeding and drug use into their analysis. Scientists studying allergies and asthma made a similar move. An initial batch of candidate genes caused much excitement, but these genes failed to consistently predict increased risk when studied in new populations (Ober 2005; Ober and Hoffjan 2006). Faced with this puzzle, scientists harkened back to David Strachan's "hygiene hypothesis," which suggested that early exposure to infections and an unhygienic environment actually conferred protection against allergies (Strachan 1989).[3] The combination of the candidate gene studies with the hygiene hypothesis presented a new, gene-environment interaction approach to understanding and studying allergies, the thought being that risk of allergies resulted from a combination of genes that influenced the immune system with environmental exposure to antigens that assaulted the immune system early in life.

Gene-environment interaction research is now a dominant paradigm in allergy science. Dozens of candidate genes have been identified that play a variety of roles in the immune system ranging from innate immunity to lung function; these genes have been combined with a variety of environmental exposures ranging from diet to air pollution (Martinez 2005; von Mutius 2009; London and Romieu 2009; Vercelli 2010). Rather than superficially reviewing this broad literature, let me focus in on two cases of interaction that have been reported, both involving the *CD14* gene. The CD14 molecule (short for "cluster of differentiation 14") is part of our innate immune system; when endotoxins in the environment find their way onto or into our bodies, CD14 binds to them in order to initiate the immune system's response to the perceived invader. That immune response is inflammatory in nature, leading to the red eyes of atopic conjunctivitis or the breathing difficulty of allergic asthma. The CD14 molecule is produced by the *CD14* gene. In 1999, Fernando Martinez and his colleagues at the University of Arizona identified a single nucleotide polymorphism (SNP) in the *CD14* gene; individuals have either a cytosine (C) or a thymine (T) at the 159th position of the *CD14* gene, meaning that we carry one of three possible variants: C/C, C/T, or T/T (Baldini et al. 1999).

In 2004, a team of scientists at the Universities of Wisconsin and Chicago combined genetic information about *CD14* with a known environmental variable from the hygiene hypothesis—pet exposure early in life—to study how these two factors affected the development of allergies. James Gern and his fellow researchers utilized the Childhood Origins of Asthma (COAST) study population, which includes almost 300 families from

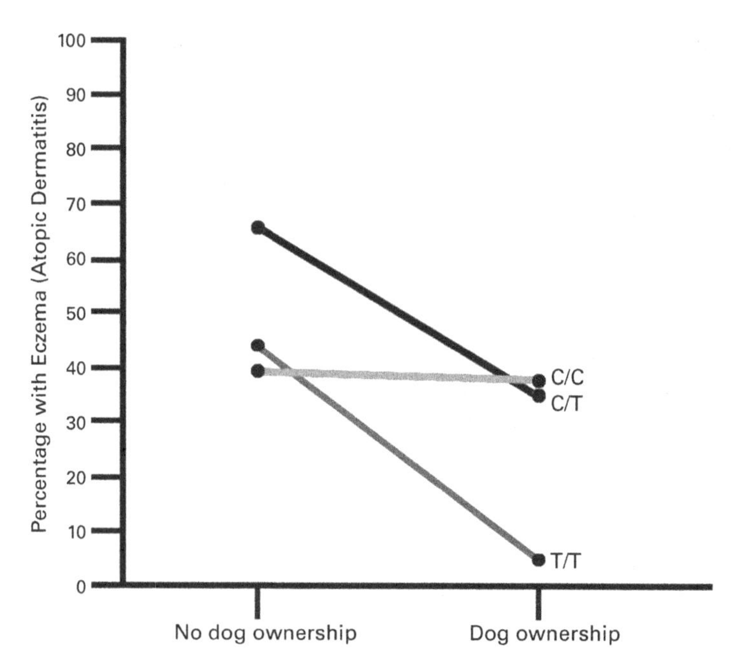

Figure 8.1
Norm of reaction graph for *CD14*, exposure to dog ownership in first year of life, and development of eczema (atopic dermatitis) in the first year of life (adapted from Gern et al. 2004, table IV).

Wisconsin who were all enrolled because at least one parent in the family suffers from asthma, allergies, or both. Gern divided up the children in these families based on (a) which variant of the *CD14* gene they carried, (b) whether they were raised in a home with a dog or a cat during their first year of life, and (c) whether the children subsequently developed atopies such as eczema. Figure 8.1 displays the results of the children's exposure to dogs. As you can see, children with either the C/T or the T/T variants of the *CD14* gene saw their risk of eczema go down quite dramatically when raised in homes with a dog present; children with the C/C variant, in contrast, saw no such benefit from dog exposure. The effect of dog exposure on the children with the T/T variant was particularly striking; over 40% of the children with the T/T variant raised in homes without dogs developed eczema, but only 5% of the children with the T/T variant raised in homes with dogs developed eczema, virtually eliminating the risk. The researchers found no such interaction effect between the *CD14* gene and cat exposure (Gern et al. 2004).[4]

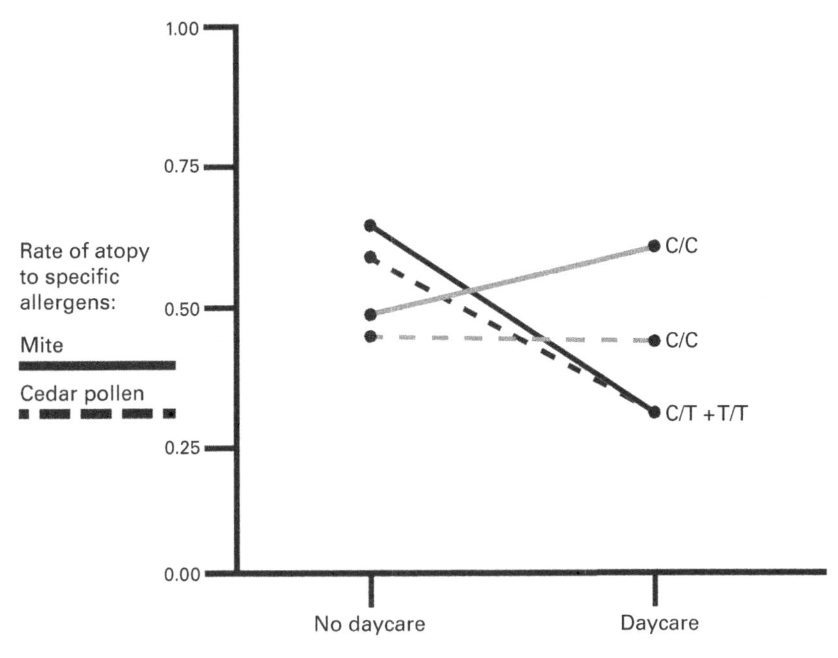

Figure 8.2
Norm of reaction graph for *CD14*, exposure to daycare in the first two years of life, and development of atopy to either mite or cedar pollen (adapted from Suzuki et al. 2009, table II).

Researchers at Chiba University in Japan found a similar phenomenon several years later. Like Gern and his team, Yoichi Suzuki and his colleagues studied the *CD14* gene, but rather than examining the SNP at the 159th position, they studied a SNP at the 550th position (another cytosine or thymine possibility). And rather than combining *CD14* with pet exposure, they combined *CD14* with daycare exposure in infancy (before age 2). Suzuki combined this genetic and environmental information from roughly 400 Japanese elementary school children and then looked at the resulting rate of atopy to various environmental antigens such as mite and cedar pollen. Suzuki's results, graphed in figure 8.2, echoed Gern's results. For children with either the C/T or T/T variants, exposure to daycare reduced the rate of mite and cedar pollen atopy (from 0.61 to 0.32 and 0.57 to 0.32, respectively); but for children with the C/C variant, exposure to daycare had no such protective effect and even slightly increased the rate of mite and cedar pollen atopy (from 0.49 to 0.59 and 0.45 to 0.46; Suzuki et al. 2009).[5]

Childhood Defiance: Parenting Styles and the *DRD4* Gene

Let me now switch from a medical condition to a behavioral one. "Externalizing behavior" refers to a cluster of childhood behavioral problems, such as defiance, aggression, and disruption. These behaviors are grouped together as externalizing behaviors because they all involve negative action directed at the child's external environment. (Behaviors such as anxiousness, withdrawal, and inhibition, in contrast, are grouped together as internalizing behaviors.) Of course, all toddlers struggle with the occasional temper tantrum, as any parent who's raised a child through the "terrible twos" can attest. But a persistent pattern of externalizing behavior early in life (roughly between the ages of 1 and 4) is unfortunately predictive of serious behavioral problems later in life—juvenile delinquency in adolescence and antisocial behavior in adulthood.

Marian Bakermans-Kranenburg and Marinus van IJzendoorn, of the University of Leiden in the Netherlands, have spent a decade studying the relationship between nature and nurture as it relates to the development of externalizing behavior. Nature, for them, has consisted in a focus on the *DRD4* gene; *DRD4* codes for a dopamine receptor, and so it figures prominently in the neurotransmission of dopamine in the brain. Within *DRD4* there is a location in the gene where a string of base pairs repeat themselves (a "variable number tandem repeat"); this repetition can happen several times or it can happen many times. Individuals with 7 or more repeats ("7+") have been distinguished from individuals with less than 7 repeats ("7–") because the longer form of the gene is associated with lower dopamine-reception efficiency (Schmidt et al. 2002). On the nurture side, Bakermans-Kranenburg and van IJzendoorn investigated the impact of different parenting styles on children. In particular, they distinguished between high-sensitivity parenting and low-sensitivity parenting. High-sensitivity parenting involves an empathy-oriented approach to discipline; when a child misbehaves (say, with a temper tantrum), a high-sensitivity parent will empathize with the child's frustration, reason with the child about the source of the frustration, and distract the child by focusing her/his attention on something else less aggravating. High-sensitivity parenting, advocates argue, leads to an increased sense of security and attachment for the child. Low-sensitivity parenting, in contrast, involves a more punishment-oriented approach to discipline; when a child misbehaves, a low-sensitivity parent, rather than empathizing, reasoning, or distracting, will be more likely to coerce the child with potential punishment as a warning against continued misbehavior. Low-sensitivity parenting, in comparison to high-sensitivity parenting, is thought to lead to insecurity and decreased attachment for the child (Ainsworth et al. 1978; van IJzendoorn 1995).

In a series of fascinating studies, Bakermans-Kranenburg and van IJzendoorn have tracked how children with different variants of the *DRD4* gene respond to different parenting styles when it comes to externalizing behavior. As a measure of externalizing behavior, they utilized the Child Behavior Checklist (CBCL). A CBCL score derives from answers that parents give to questions about their child; for example, in answer to a category like "argues a lot" or "disobedient at school," parents can answer "often true" (worth 2 points), "sometimes true" (worth 1 point), or "not true" (worth 0 points). So a higher score on the CBCL reflects increased signs of externalizing behavior in the child. In Bakermans-Kranenburg and van IJzendoorn's first study, they identified a striking case of an interaction that changes rank. As figure 8.3 shows, children with the 7+ repeat of *DRD4* showed a drastic decrease in externalizing behavior when exposed to high-sensitivity parenting (versus low-sensitivity parenting), cutting their

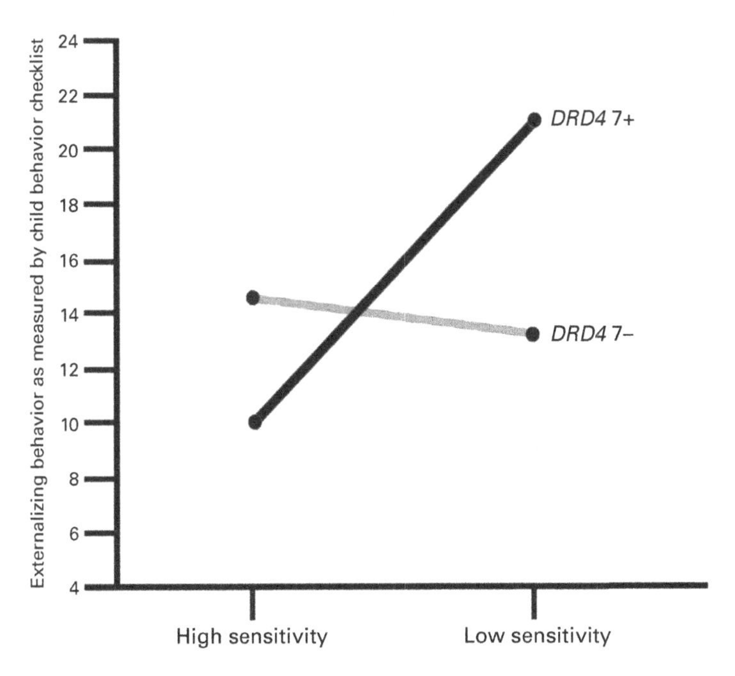

Figure 8.3
Norm of reaction graph for *DRD4*, exposure to high- vs. low-sensitivity parenting, and externalizing behavior as measured by the Child Behavior Checklist (reproduced from Bakermans-Kranenburg and van IJzendoorn 2006, figure 1).

CBCL score in half. 7– children, in contrast, saw no such decrease when exposed to high-sensitivity parenting; in fact, 7– children's CBCL score slightly increased when exposed to high-sensitivity parenting (Bakermans-Kranenburg and van IJzendoorn 2006).

With the results of that study in place, Bakermans-Kranenburg and van IJzendoorn next wondered how children with different variants of *DRD4* would respond to an intervention designed to increase parental sensitivity. They started by identifying high-risk children—one- to three-year-olds who already scored in the 75th percentile of the CBCL. A total of 157 children and their parents were enrolled, with half acting as a control group and the other half receiving the intervention. The intervention involved a series of six visits (each lasting ~1.5 hours) over eight months; during these visits, the intervener would video mother-child interactions and then play video back to the mother in order to emphasize times where the parent was successfully engaging in high-sensitivity parenting and times where the parent could improve in that endeavor. To judge the effectiveness of the intervention, CBCL scores were collected at the screening phase (when the children and families were first identified), at a pretest phase (four months later and just before the children and families were randomly assigned to the intervention or control groups), at a posttest phase (several months after the intervention was completed), and finally at a follow-up phase (one year after the posttest). There was a slight increase in CBCL scores from the screening to the pretest phases for all groups followed by a decline in CBCL scores from the posttest to the follow-up phases for all groups. However, children with different variants of the *DRD4* gene responded to this intervention differently. Children with the 7+ repeat responded quite well to the intervention; they had the highest CBCL scores without the intervention but the lowest CBCL scores with the intervention. 7– children, however, saw the exact opposite effect; their CBCL scores slightly increased when exposed to the parental sensitivity intervention (Bakermans-Kranenburg et al. 2008b, figure 1).[6] More recently, Bakermans-Kranenburg and van IJzendoorn have tracked the differential impact of *DRD4* and parenting styles beyond the toddler years to later childhood. But rather than tracking the negative impact evinced by externalizing behavior, they tracked the positive impact reflected in altruistic behavior. This time, 91 7-year-old children were enrolled. The children were first given 10 €0.20 coins purportedly for their participation in the study; the children were then shown a two-minute UNICEF promotional film about a child in a poor country, the ending of which asked the children to donate by placing money in a box situated in the same room as the children watching the film. When the film ended, a member of the research team entered the room and asked the children if they would like to donate money to UNICEF. Again, Bakermans-Kranenburg and van IJzendoorn linked up differences in the children's *DRD4* gene and differences in the children's exposure to different parenting styles (this time distinguished based on secure versus insecure attachment parenting) to differences in donations. And again they found the interaction result that changed in rank. For children with the 7+ repeat of *DRD4*, exposure to secure parenting coincided with increased donations. But for the 7– children, exposure to secure parenting slightly decreased their donating behavior (Bakermans-Kranenburg and van IJzendoorn 2011).[7]

Taking Stock of Differential Susceptibilities

The three cases above—*CD14* and dog exposure, *CD14* and daycare exposure, and *DRD4* and parenting style exposure—all displayed an interaction effect between a particular gene and the environment a child was exposed to early in life. What's more, the interaction effect in all three cases was so extreme that the norms of reaction changed rank from one environment to the next, reflecting a differential susceptibility to allergies and externalizing behavior (Belsky, Bakermans-Kranenburg, and van IJzendoorn 2007; Ellis et al. 2011). What was most striking about these cases of differential susceptibility was how differently the children responded to the environmental exposures depending on which gene they carried. In all three cases, children with one variant of a gene showed a marked decrease in risk when exposed to dogs, daycare, and empathy-oriented discipline. But children with the other variant experienced no such benefit from those same exposures. And, in fact, they may have even experienced the *opposite* effect. Across multiple studies, the 7– children in Bakermans-Kranenburg and van IJzendoorn's research increased their externalizing behavior when exposed to high-sensitivity parenting. Likewise, children with the C/C variant of *CD14* in Suzuki's study who were exposed to daycare, rather than decreasing their rate of atopy, increased it. Now, importantly, these increases were not statistically significant, meaning that they could just as well have been the result of chance. But then all that means is that, based on the current information, high-sensitivity parenting *might not* increase a 7– child's externalizing behavior and that daycare *might not* increase a C/C child's risk of allergy. For parents, the difference between "likely to decrease risk" and "might not increase risk" could be influential when making decisions about the family pet, childcare, or how to handle the next temper tantrum.

The "$1,000 Genome" and Demand for It

Might tests for genes such as *CD14* and *DRD4* become widely available? And, if so, might parents want and acquire that genetic information about their children? If the answer is "no" to either of these questions, then there's no use in proceeding with a bioethical analysis of the advantages and disadvantages of parents utilizing this information. But there is actually good reason to think the answer is "yes" to both questions.

When the first genetic tests were developed, they tested specific stretches of DNA associated with specific diseases. The idea that an individual's entire genome would be sequenced was ridiculous because the process was simply cost-prohibitive. The first human genome took ten years to sequence and cost $3 billion dollars, but as sequencing technologies improved it suddenly became possible to provide a personalized genomic picture of individuals ranging from Nobel laureate James Watson to Archbishop Desmond Tutu (Wheeler et al. 2008; Schuster et al. 2010). Information about one's genome, the promise went, would allow for personalized medicine, where everything from diet to prescriptions could be tailored to one's individual genes. Now the cost of whole-genome sequencing is down to several thousand dollars and takes roughly one week. Biotech companies are promising the "$1,000" genome, which will take only hours to sequence, in the very near future (Hayden 2009). Of course, it remains to be seen how long it will take biotech companies to deliver on this promise (where the $1,000 price tag includes both sequencing the genome and interpreting the results). Still, it's safe to say that the race is on and the trend is heading toward a more and more affordable product. This is why the answer to the question about the availability of genetic tests for *CD14* and *DRD4* is "yes." It's not "yes" because specific genetic tests are being developed to provide genetic information about those specific genes; rather, it's "yes" because that genetic information will simply be a byproduct of whole-genome sequencing, which is becoming increasingly affordable for anyone interested in learning that genetic information about themselves ... or their children.[8]

Would parents, though, really want the entire sequence of their child's genome? Again, recent developments suggest the answer is "yes." A team of researchers at Case Western University and the University of Michigan asked a nationally representative sample of over 1,500 parents about their interest in whole-genome sequencing for a hypothetical future newborn. The parents were presented with one of two possible scenarios—either the whole-genome sequencing was offered as part of a state's preexisting newborn screening program, or the whole-genome sequencing was offered in a pediatrician's office.[9] The parents were then asked how interested they would be in getting the whole-genome sequence of their newborn, with answer options ranging from definitely not interested, to not interested, to somewhat interested, to definitely interested. For the parents presented with whole-genome sequencing as part of the state's newborn screening program, 74% expressed interest in the opportunity (36% were definitely interested, 38% were somewhat interested, 18% were not interested, and 8% were definitely not interested). For the parents presented with the whole-genome sequencing in the pediatrician's office, 70% expressed interest in the opportunity (31% were definitely interested, 39% were somewhat interested, 22% were not interested, and 8% were definitely not interested). When asked to identify the factors that affected their decision, the majority of these parents selected "preventing or decreasing my child's chances of developing disease" and "choosing medical treatments that might be more effective for my child" as very important (Goldenberg et al. 2014).[10]

This is all to say that there is good reason to think genetic information about genes such *CD14* and *DRD4* will soon become affordable for a wide population via whole-genome sequencing, and that parents will seriously consider acquiring this genomic information about their children should the opportunity present itself. Before considering the costs and benefits of this possibility, though, several clarifications are in order. First, there is a disconnect between the scenarios presented above to the parents about sequencing their hypothetical newborn and the scenario I am considering here as a possibility for the near future. That disconnect is money. Whole-genome sequencing is predicted to drop to $1,000 soon, but that is still $1,000 more than the whole-genome sequencing offered to the parents in the study above. In their scenarios (either the state newborn screening program or the pediatrician's office), whole-genome sequencing was offered for free. Were those same parents offered whole-genome sequencing for their hypothetical newborn at a hypothetical cost of $1,000, the percentage of interested buyers would surely drop below 70%–74%. So we can think of the 70%–74% of interested parents in that study as currently representing a ceiling of interest that would decrease as the cost of the service increased. Still, roughly 4 million children are born every year in the United States, so even if that 70%–74% of interested parents were cut down to 35%, or 10%, or even just 5% because of the price tag, we're still talking about hundreds of thousands of children born every year in the United States with parents interested and willing to pay for whole-genome sequencing. And, after all, parents can easily spend $1,000 on a child's new crib.

Might not those same parents spend that amount to gather genomic information about their child sleeping in the crib?

The second clarification concerns replications of the *CD14*/dogs, *CD14*/daycare, and *DRD4*/parenting results. Those results are all quite new, and they have not received nearly the scrutiny that the gene-environment interaction results by Moffitt and Caspi have received (discussed in chapter 4). As a result, it is still too early to say with any certainty that these results are validated. To refer back to chapter 6, variation-partitioning studies will need to confirm that the interaction effect is in fact real by undertaking replications in new populations, and mechanism-elucidation studies will need to show how that purported interaction effect percolates up from the molecular/cellular levels to the medical/behavioral levels by undertaking experimental investigations. These studies will also have to confirm the apparent trend for children with one genetic variant to respond favorably to an environmental exposure while children with the other variant respond negatively to that same exposure, determining whether or not the unfavorable response is real or just a chance artifact of the early studies.

Shall we wait for these results to be replicated before proceeding with any bioethical analysis of their implications? That would be unwise. One of the functions of the bioethicist is to get out ahead of the science in order to provide an ethical evaluation of some technological development before that technology actually falls in our laps. The history of empirical evidence for interaction reviewed in chapter 6 showed that, while interaction is no guarantee in any particular case, it has consistently turned up for a range of gene-environment combinations relating to a range of human traits. My focus in this chapter is on those gene-environment combinations where the environmental variable involves a childhood exposure which parents have been deciding upon for generations without any consideration of the child's genes. So even if it is not these particular gene-environment combinations, history suggests that some validated gene-environment interaction results will arise that involve parental decisions as the shaper of the environmental variable. What, then, would it look like for genetic information to suddenly start influencing those parental decisions?

A "Genetic Guide to Parenting"?

Imagine the following scenario playing out in a pediatrician's office: a young couple has just welcomed a second child into their family. They already have an older son—Vince, to whom they are now adding the newborn—Jeff. At Jeff's first visit to the pediatrician, the physician asks the parents if they would like to pay $1,000 in order to sequence Jeff's entire genome. When they ask what such genetic information will provide, the pediatrician explains that the whole-genome sequence can influence Jeff's care in several ways. For instance, it might turn up rare genetic diseases with known medical treatments; learning of that genetic information quickly allows for immediately responding with the medical treatments. It might also identify cases of gene-environment interaction, where decisions that the parents will make anyway about Jeff's environment can shape the influence of genes on traits ranging from allergies to defiant behavior. The parents themselves both suffer from hay fever and eczema and so does their three-year-old Vince, so the thought that something could be done to prevent the development of these conditions in Jeff is tempting. What are these decisions they will make anyway, the parents ask, that could shape the influence of genes on allergies? Dogs and daycare, the pediatrician responds. Depending on which variant of the *CD14* gene Jeff carries, being raised in a home with a dog could either significantly reduce his risk of developing eczema or have no effect at all, and being sent to daycare could either significantly reduce his risk of developing certain allergies or could potentially increase his risk of developing those same allergies. The parents own a cat, but the pediatrician explains that cats don't seem to have the same effect; the parents have never owned a dog and had no immediate plan to add one to the family. Turning to daycare, the parents had not yet made a decision about it, waiting to see what it would cost to send two children. But the prospect of being able to actively shape Jeff's risk of developing allergies by making decisions about a pet or childcare is attractive; they spend several hundred dollars each year on treatments for their own and Vince's allergies, so the thought that $1,000 now could prevent the development of such allergies in Jeff sounds like a good bet. They proceed with the whole-genome sequencing and add that information to Jeff's medical record, agreeing with their pediatrician that the genomic information will be consulted when it's relevant to a decision about Jeff's care. For now, because allergies are the focus, they learn of their child's *CD14* status: at the 159th position, Jeff carries the T/T variant (suggesting that a dog will reduce his risk of eczema), and at the 550th position, Jeff carries the C/T variant (suggesting that daycare will reduce his risk of certain allergies).

At a one-year check-up, the pediatrician asks about how Jeff is doing, especially with regard to the family history of allergies. While not allergy-free, the parents report that Jeff is much better off than either his brother or his parents. He doesn't have eczema like his brother, but he does suffer from some of the same hay fever as the rest of the family. The lack of eczema they attribute to Dander—their two-year-old dog, which they added soon after learning of Jeff's likelihood to benefit from dog exposure. Vince, their older son, now takes an additional allergy medication to prevent his allergy to Dander; still, it seems to have done the trick for Jeff. But Jeff's hay fever, the parents worry, may be the result of their decision not to send him to daycare. For financial reasons, it makes more sense to have one parent stay at home with the kids rather than send them to costly daycare. Might daycare have prevented Jeff's hay fever, the parents ask? It's hard to say, their pediatrician replies; these gene-environment interactions are probabilistic in nature, not deterministic, so Jeff could very well have gone to daycare and still developed hay fever. The parents nod, but seem unconvinced.

One year later, at Jeff's two-year check-up, the conversation between pediatrician and parents shifts from medical issues to behavioral ones. The parents complain that Jeff is becoming increasingly difficult to parent. Temper tantrums are common, as is his habit of meeting parental demands with aggressive opposition. The parents have been trying to crack down on this by increasing the severity of punishment (which worked well with Vince), but this disciplinary strategy only seems to escalate the defiance in Jeff. The pediatrician suggests that they refer back to Jeff's whole-genome sequence and find out which variant of the *DRD4* gene Jeff carries. *DRD4*, the pediatrician notes, has been implicated in a gene-environment interaction with parenting style. If Jeff has the long version of the gene, then an empathy-oriented approach to discipline will likely be more successful than the punishment-oriented approach that the parents have been using (and used on Vince). The parents consider the apparent effectiveness with which Dander prevented Jeff's eczema, so incorporating another gene-environment interaction into their parenting decisions makes sense. They ask to learn of Jeff's *DRD4* status, and it turns out that he does indeed carry the 7+ long variant. The parents, worried that they may have been mistreating Jeff in light of his genetic status, promise to switch over to an empathy-oriented approach to disciplining their younger son.

Predictive and Preventive Genetic Testing in Children

Predictive and preventive genetic testing in newborns and children is nothing new. Nor is the ethical analysis of when such testing is and is not appropriate. The medical model involves a specific search for a specific gene (or gene product) to determine whether or not a specific disease is likely to develop. If the gene in question is present, then learning this information quickly allows for responding with medical treatments designed to prevent the disease from developing (or, perhaps, slowing it down). The ethical analysis consists in weighing the benefits of medical treatment against the various social, familial, and psychological costs of acquiring that genetic information about the child (Andrews et al. 1994; Wertz, Fanos, and Reilly 1994; Hoffmann and Wulfsberg 1995).

An international consensus has emerged surrounding this analysis (Borry et al. 2006). For childhood-onset diseases with effective medical treatments available, genetic testing should take place. For example, children in the United States are screened at birth for phenylketonuria (PKU), a metabolic disorder wherein (for genetic reasons) a child cannot metabolize phenylalanine, an amino acid common in food and drink. If the child with PKU ingests phenylalanine, then the amino acid will build up in the child's body and ultimately lead to debilitating mental handicap and seizures. Fortunately for children with PKU, there is a very effective medical treatment available; if the child is put on a phenylalanine-free diet, then these disabilities can be avoided and the child will not suffer from the symptoms of the genetic disorder. Another example is familial adenomatous polyposis (FAP). A child who inherits FAP (via a mutated *APC* gene) will begin forming thousands of polyps on her/his colon starting in early adolescence; if these polyps are left untreated, they will develop into malignant colon cancer, but if the polyps are monitored with regular colonoscopies, then a surgery to remove the colon can be performed before cancer forms.

PKU and FAP are often thought of as "genetic conditions," but with medical treatments available, they are better thought of as presenting cases of gene-environment interaction. Exposure to the medical treatment is the environmental variable. Without exposure to the medical treatment, individuals with the various genes would develop (or would be more likely to develop) the diseases in question, but with exposure to the medical treatment, individuals with the various genes can eliminate (or reduce) their risk of developing the diseases. The situation is graphed in figure 8.4. For individuals with the mutated *APC* gene associated with familial adenomatous polyposis, the risk of cancer goes from very high to much lower by moving from an environment without a preventive colectomy to moving to an environment with a preventive colectomy. For individuals with a functioning copy of the *APC* gene, though, the risk of cancer was lower to begin with and so only drops down slightly more by removing the colon.

In one sense then, familial adenomatous polyposis is similar to the cases of allergies and externalizing behavior discussed earlier. With exposure to

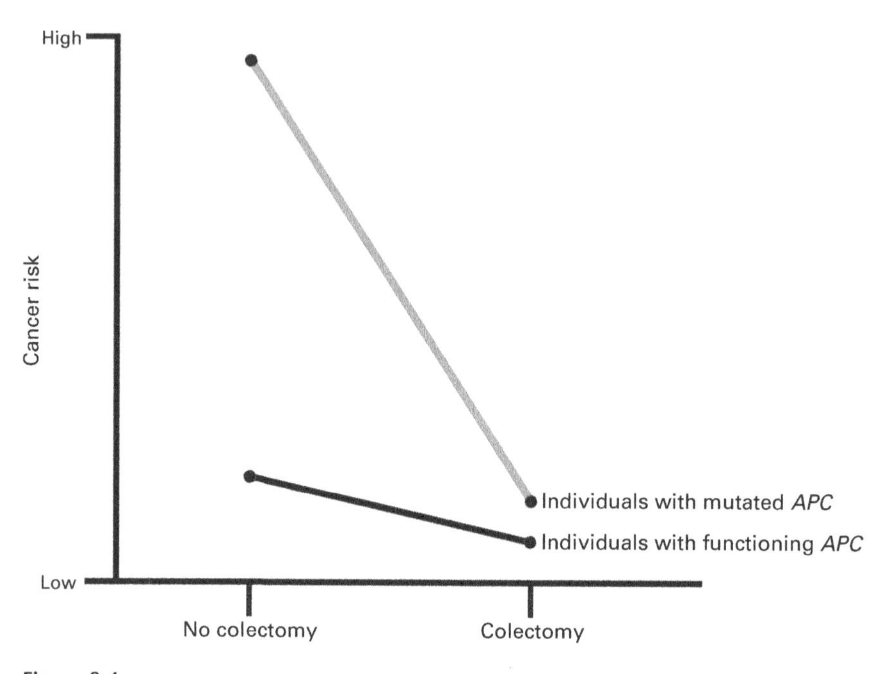

Figure 8.4
Norm of reaction graph for *APC*, exposure to a preventive colectomy, and risk of developing cancer.

the right environmental variable, different genetic groups respond differently. In another sense though, FAP is much different from allergies and externalizing behavior. The difference has to do with what counts as the "right environmental variable." The medical treatment for FAP is an invasive surgery. For PKU, it is a highly specialized diet. Were it not for the FAP or PKU, parents would not have to decide upon removing their child's colon or placing their child on a phenylalanine-free diet. These decisions about medical treatment are faced only because of the medical diagnosis. But with the particular cases of gene-environment interaction concerning allergies and externalizing behavior, the situation is different. Right now, somewhere a parent is deciding how to respond to their toddler's temper tantrum, and they are making that decision with no regard to the contents of their child's medical record. They're just parenting. Right now, somewhere a couple is deciding whether or not to send their child to daycare, just as their parents did before them. It's a decision about finances, about career trajectories, and about philosophies of parenting. Parents, in short, have been making decisions about such environmental exposures of their child for generations, but the results of a genetic test had nothing to do with them. Now, by situating those environmental exposures in gene-environment interactions, suddenly the results of a genetic test could be relevant. And by linking that gene-environment interaction to a medical or behavioral trait, suddenly deciding upon that environmental exposure could be interpreted as deciding upon treatment for that trait. This incorporation of parental decisions into gene-environment interactions associated with child development stands to alter how parents think about a range of issues: the rationale for such decisions, the responsibility for the outcome, and the nature of the relationship between decision and outcome. This alteration, I hope to show, will be for both better and for worse.

Parental Decision Making: Empowered ... and Medicalized

When parental decisions about the environmental exposures of children can shape gene-environment interactions associated with various medical or behavioral traits, the very concept of "medical treatment" is expanded. A decision about whether to add a dog or a cat to the family or a decision about how to discipline a child can be reconceived as treatment for a child's genes—the equivalent of surgery for FAP or a special diet for PKU. On the one hand, the reconceptualization of these decisions could be empowering for parents. Traditionally, medical treatments for genetic conditions are just that—medical. Think about the parents whose child has been diagnosed with FAP. A series of regular colonoscopies will be scheduled, eventually followed by a colectomy. These procedures will take place in a health care facility; they will be performed by medical specialists; they will involve medical care in preparation for and then in response to the procedures. Adopting a dog, in contrast, requires no such oversight from the world of medicine. It is a decision that parents can make for their family simply by going down to the local animal shelter. And yet, given the right genetic variant in their child, that addition of a dog to the family can be as much a treatment for the child's allergy risk as any prescription drug from a pediatrician. Reconceiving everyday parental decisions as treatments endows those decisions with the power of medicine, and putting that power directly into the hands of parents as opposed to medical professionals empowers parents when they are able to make those decisions.

At the same time, reconceiving certain parental decisions as treatments stands to medicalize those decisions, fundamentally altering how parents think about the rationale for them. Parents normally add a dog to the family in order to teach their child about responsibility and companionship and to give a loving dog a loving home. But as a preventive treatment for allergies, the dog becomes reconceptualized as a dander-dispenser. Likewise, parents normally discipline their child in order to teach the child a lesson about what is appropriate and inappropriate behavior. But as a preventive treatment for externalizing behavior, the discipline becomes reconceptualized as a treatment for the child's *DRD4* gene. This alternative rationale for parental decisions turns those decisions into medical means to medical ends (Conrad 2007; Conrad, Mackie, and Mehrotra 2010; Parens 2013). The act of reconceptualization is not inherently bad, but it does have consequences. Will parents with two children who have two different versions of *DRD4* discipline those children differently based on that genetic difference? Consider Vince, the older brother in the scenario I described above. His parents took a punishment-oriented approach to him when he misbehaved, but now they are planning to take a very different empathy-oriented approach to his younger brother's misbehavior. Conceived of as treatment in response to a gene-environment interaction, this difference in style of discipline makes sense. But that medical rationale is unlikely to assuage Vince, who will see an injustice in how he is met with punishment when Jeff is met with empathy; perhaps Vince will take that resentment out on Jeff. The worry here is that a "genetic guide to parenting" will trump other guides to parenting (such as being fair when it comes to discipline) by virtue of the perceived importance of genes in shaping who we are (Dar-Nimrod and Heine 2011).

Parental Responsibility: Taking It … and Bearing It

Empowered parental decision making is joined by validation of parenting when those decisions have positive outcomes. Take the addition of the dog to Jeff's environment in the scenario above. The parents chose to add Dander to the family; when Jeff failed to develop eczema, the parents could look to the decision they made and feel good about it. They did not have to thank a surgeon for removing Jeff's colon and preventing colon cancer. It was their action that contributed to the positive result. This is the outcome side of empowered parental decision making—it allows parents to take responsibility for a positive result.

At the same time, the ability to take responsibility brings with it the burden of bearing responsibility. When a child inherits a genetic condition from her/his parents, there is a danger that the parents will feel guilty for supplying their child with a faulty gene (American Society of Human Genetics Board of Directors and American College of Medical Genetics Board of Directors 1995; Aatre and Day 2011). But a pediatrician or genetic counselor can counter this guilt by explaining to the parents that they need not feel responsible for this result; which genes a child receives is an inherently probabilistic affair, so (short of in vitro fertilization joined by preimplantation genetic diagnosis) there is simply no way the parents can definitively prevent the inheritance of those rare genes. The environmental variable is different, though, since the parents can do something about it. The parents of Jeff in the scenario above opted out of sending Jeff to daycare for financial reasons. Had they sent him, might he not have developed hay fever? As the pediatrician explained, there's really no way to know since the decision would have only reduced risk, not eliminated it. But notice that the pediatrician or genetic counselor's explanation here is quite different from the one given to the parents who have a child with a rare genetic disorder. With a child who has a rare genetic condition, there is a sense that the parents simply must deal with this hand that luck has dealt them and their child. But with a child who has a condition associated with a gene-environment interaction the environment of which is under the parents' control, there is a sense that the parents could have shaped the environment differently so as to alter the likelihood of the result. This is the outcome side of medicalized parental decision making—it challenges the parents to take responsibility for a negative result.

Determinism: Undermined … and Magnified

Incorporating gene-environment interactions into parental decision making will undermine common ideas about genetic determinism. This would be a good thing. The idea that genes deterministically decide who we are or are to become is both old and persistently all too common (Oyama 1985, 2000; Allen 1997; Kaplan 2000; Fox Keller 2000; Robert 2004; Griffiths 2006; Moore 2008b; Stotz 2008; Lickliter and Honeycutt in press). Perhaps this should not be surprising. Traditional genetics education uses Mendel's results about pea plants (e.g., round vs. wrinkled seeds, long vs. short stems, yellow vs. green pods) as the paradigm of gene action. For most of the public, the next time they think about genes is when they are having a child and speaking with a pediatrician or a genetic counselor about their child's risks of various genetic conditions. Both experiences reinforce a very deterministic, gene-centered perspective concerning how genes work and how traits develop. With cases of gene-environment interaction, though, the effects of a gene are inextricably linked to the effects of the environment. A team of scholars at the University of Leeds is trying to combat the educational side of this equation; Gregory Radick, Annie Jamieson, and Jenny Lewis are developing an alternative genetics pedagogy that uses cases of gene-environment interaction as the paradigm of gene-action and trait-development (Jamieson and Radick 2013). It will be intriguing to follow the results of this alternative approach to genetics education. Introducing cases of gene-environment interaction to parents processing medical information about their child would combat the other side of the equation; if parents heard about the interactive relationship between *CD14* and dogs as it relates to allergies alongside the genetic tests for genetic conditions like PKU, they would be less likely to internalize a deterministic, gene-centered perspective concerning how genes influence the development of their child's traits.

At the same time, these cases of gene-environment interaction could magnify perceptions of determinism, albeit of a different sort. That is, while simple genetic determinism may be undermined, a more powerful sense of gene-environment determinism could replace it. Consider Jeff again from the scenario above. The results of the whole-genome sequence allowed Jeff's pediatrician to inform his parents that Jeff has the 7+ form of the *DRD4* gene. Bakermans-Kranenburg and van IJzendoorn's research associated the 7+ variant with externalizing behavior in children but only in coordination with low-sensitivity parenting. On the one hand, this clearly undercuts a perspective of genetic determinism as it relates *DRD4* to externalizing behavior. On the other hand, a child with the 7+ form of the gene who has also been exposed to low-sensitivity parenting could be conceived of as being hit with a double-dose of risk for externalizing behavior. Jeff's parents have promised to switch over to a high-sensitivity approach to discipline, but might not every subsequent temper tantrum by Jeff be interpreted differently now—as a genetic condition the parents have been aggravating with their incorrect environmental response to that condition? By incorporating both a genetic risk and an environmental risk, the worry is that exposure to both will magnify a deterministic perspective of trait-development, even if that perceived determinism is more complex than simple genetic determinism.

Genetic Testing for Interaction: What Is (and Is Not) to Come

I opened part III of this book by asking, "What does the future hold for the interaction of nature and nurture?" I hope to have conveyed in the last two chapters what I think that future does and does not hold. Despite widespread attention to the case of interaction between *MAOA*, childhood maltreatment, and antisocial behavior, the future does not hold preimplantation genetic diagnosis or statewide newborn screening for *MAOA*. These interventions are both misguided and unlikely because (a) they mischaracterize low-MAOA individuals as genetically predisposed to violence, and because (b) they overlook a much simpler solution to combating antisocial behavior—targeting the environmental variable by decreasing the prevalence of childhood maltreatment for all children regardless of their genes.

The cases of interaction discussed in this chapter, in contrast, are much more viable candidates for gene-guided interventions precisely because the environmental variables discussed here are not obviously objectionable. Unlike childhood maltreatment, there isn't an obviously right answer when it comes to parental decisions about pets, daycare, and disciplinary style. Those decisions are made by parents all the time, and a variety of factors influence those decisions. Between the falling costs of whole-genome sequencing and parental demands for it, it seems likely that genetic information about the child could become one of the factors influencing those decisions. This won't necessarily be a bad thing. To the extent that parents can utilize information about particular gene-environment interactions to tailor their children's early life exposures and experiences so as to reduce the risk of diseases and disorders, this will be an empowering experience for parents, which will simultaneously combat widespread misconceptions about genetic determinism. Still, parents will have to enter genetic information into these decisions with great care. If parents think the "genetic guide to parenting" bears more weight than other factors, there is a genuine risk of medicalizing those decisions and replacing genetic determinism with gene-environment determinism. Efforts at translating gene-environment interaction research results into clinical interventions that can guide parents making decisions about their child's environment will be challenged with balancing these costs with these benefits.

Notes

Chapter 1

1. This twin method is itself subject to criticism, which I will discuss in subsequent chapters. All we need to know at this point is that it convinced Galton that nature won the debate.

2. A debate played out in the pages of the *American Journal of Epidemiology* in the late 1970s and early 1980s (Walter and Holford 1978; Rothman 1978; Kupper and Hogan 1978; Blot and Day 1979; Saracci 1980; Rothman, Greenland, and Walker 1980).

Chapter 2

1. This chapter will focus on eugenics in the United States and the United Kingdom since the debates over interaction discussed in this chapter and the next played out there. Extensive historical research has been done on the rise, proliferation, and demise of eugenics in these two countries (see, for example, Blacker 1952; Ludmerer 1972; Soloway 1990; Mazumdar 1992; Kevles 1995; Paul 1995, 1998; Pernick 1996; Rosen 2004; Bashford and Levine 2010). For studies of eugenics in other countries, see Adams (1990) on Germany, France, Brazil, and Russia; Stepan (1991) on Latin America; Broberg and Roll-Hansen (2005) on Scandinavia; Bashford and Levine (2010, chapters 11 through 31) on a range of nations; and Harris-Zsovan (2010) on Canada.

2. "Feebleminded" was a technical term for eugenicists. Feebleminded persons suffered from mental defect, but they were on the high-functioning end of the mental defect spectrum, in contrast to the "idiot" and the "imbecile" at the other end.

3. Box's biography of Fisher remains the most complete work on his life. Shorter treatments can be found in Yates and Mather (1963) and Mahalanobis (1964).

4. For historical reflections on the biometric-Mendelian debate, see Froggatt and Nevin (1971), Cock (1973), Norton (1973, 1975), Darden (1977), Roll-Hansen (1980), MacKenzie (1981a), Olby (1988), and Provine (2001).

5. Fisher's presentation is reprinted as Appendix iii in Norton and Pearson (1976, 155–162).

6. Biometrician George Udny Yule (1902), sixteen years earlier, had considered the same problem and argued that the Mendelian principles of inheritance could be seen as a special case of the biometric law of ancestral heredity (Tabery 2004a); Fisher, in contrast to Yule, took the reductive relationship between the Mendelian principles and the biometric law of ancestral heredity in the opposite direction (Sarkar 1998, 106).

7. For discussions of the significance of Fisher (1918), see Yates and Mather (1963), Mather (1964), Moran and Smith (1966), Kempthorne (1974), Thompson (1990), and Provine (2001).

8. Mackenzie was an assistant to Fisher at Rothamsted; they published several papers together while Fisher was at Rothamsted (Fisher, Thornton, and Mackenzie 1922; Fisher and Mackenzie 1922, 1923). I am grateful to John Gower for information on Winifred Mackenzie.

9. Hogben wrote his autobiography, *Look Back with Laughter* (*LBL*), in the early 1970s. G. P. Wells (H. G. Wells' son) drew on much of this to write his Royal Society biographical memoir of Hogben (Wells 1978). Wells also edited Hogben's *LBL* with an eye toward publishing it in the late 1970s but could not succeed in the endeavor (Tabery 2006). More recently, Hogben's son and daughter-in-law published a heavily edited version of *LBL* under the title *Lancelot Hogben, Scientific Humanist: An Unauthorized Autobiography* (Hogben 1998). More limited biographies of Hogben can be found in Werskey (1978), Mazumdar (1992), Kevles (1995), Sarkar (1996), Gurdon and Hopwood (2000), and Tabery (2006). For the purposes of this chapter, biographical references will be made to Hogben (1998); when I cite material that was edited out of Hogben (1998), references will be to Wells' edited version of *LBL* held at the University of Birmingham Library (listed as A.9 and A.10 of the Hogben Papers).

10. Wells wrote a follow-up essay to his biographical memoir entitled "Father and Son" (A.38) that detailed the Hogben family along with Thomas Hogben's influences on his son; however, Wells could not convince the Royal Society to publish the sequel (Wells n.d.).

11. Darlington to Wells, 6 June 1976, Lancelot Hogben Papers (A.44), University of Birmingham Library.

12. Fisher to Leonard Darwin, 3 December 1929; Darwin to Fisher, 7 December 1929. R. A. Fisher Papers (available online at http://www.library.adelaide.edu.au/digitised/fisher/).

13. "Full-fledged" because Hogben had addressed critically the science of eugenics to a limited extent prior to *Genetic Principles* in earlier book chapters and lectures (Hogben 1927, 1931a, 1931b).

14. For other prominent reviews of *Genetic Principles* by Hogben's contemporaries, see Huxley (1932) and Haldane (1932).

15. Fisher was not the only contemporary of Hogben's put off by this demand for an experiment-first approach to science. Hogben applied the same criticism to his colleagues at the LSE and rubbed many of the theoretically oriented social scientists there, such as Friedrich Hayek, the wrong way. This clash ultimately proved insolvable, and Hogben left the LSE in 1937 (Renwick 2013).

16. Hogben to Fisher, 17 February 1933, R. A. Fisher Papers (available online at http://www.library.adelaide.edu.au/digitised/fisher/).

17. Fisher to Hogben, 18 February 1933, R. A. Fisher Papers (available online at http://www.library.adelaide.edu.au/digitised/fisher/).

18. Hogben to Fisher, 23 February 1933, R. A. Fisher Papers (Series I, Hogben, L.), University of Adelaide.

19. Hogben was not the first to emphasize development in contrast to nature versus nurture. Other authors had already pointed to the developmental process to attack a strict divide between innate and acquired characteristics (Carmichael 1925; Krout 1931). Hogben was the first, though, to apply the move to statistical efforts at partitioning variation, and so he was the first to link up the developmental process with the interaction of nature and nurture.

20. This was not the first time Hogben revealed his philosophical indebtedness to Russell. Hogben's *The Nature of Living Matter* (1931a), a mechanistic critique of vitalism, was dedicated to Russell as well.

21. Fisher to Hogben, 25 February, 1933, R. A. Fisher Papers (Series I, Hogben, L.), University of Adelaide.

22. Fisher to Roberts, 18 January 1935, quoted in Bennett (1983, 260). Haldane did subsequently take up the issue of interaction between nature and nurture and apply it critically to eugenics in a manner similar to Hogben (Haldane 1936, 1946).

23. These references were not included in the version of his essay that appeared in *Nature and Nurture* (1933a). There, Hogben only wrote, "There is no reason to multiply instances in order to show the need for extreme care in formulating the problem of nature and nurture in quantitative terms" (ibid., 97).

24. In earlier publications (Tabery 2007a, 2008, 2009a, 2009b), I referred to Fisher's variation-partitioning approach as the "biometric research tradition" and Hogben's mechanism-elucidation approach as the "developmental research tradition." These labels, however, created confusion. On Fisher's side, the "biometric research tradition" suggested to some that I was talking about any application of statistics to biology, but that was a much wider domain than I intended to characterize. On Hogben's side, the "developmental research tradition" suggested to some that I was

talking only about developmental biologists, but that was a much too narrow domain than I intended to characterize. As a result, my hope is that making these titles more explicit in terms of the approaches they take to explanation (either partitioning variation or elucidating mechanisms) will eliminate that earlier confusion.

25. There is an interesting connection to be drawn here between Fisher's debate with Hogben and Fisher's more famous debate with Sewall Wright. Both concerned interaction—heredity-environment interaction in the case of Hogben, and gene-gene interaction (or epistasis) in the case of Wright. Fisher's responses to Hogben and Wright were remarkably similar; in both cases, he downplayed the significance of interaction in favor of additive genetic effects that his statistics were better equipped at measuring. For historical treatments of the Fisher-Wright debate, see Hodge (1992), Provine (1992), Skipper (2002, 2009), and Plutynski (2005).

26. This was a point Hogben made again twenty years later when he criticized the "tyranny of averages" in medical research (Hogben and Sim 1953). On the connection between this 1953 publication and Hogben's earlier debate with Fisher, see Tabery 2011.

27. Several philosophers of science and scientists have made distinctions along similar lines, distinguishing, for example, how questions from how-much questions or individual development from individual differences. I will discuss and engage the contributions from these individuals in chapter 5, where the task is bridging the explanatory divide between the two approaches.

28. R. A. Fisher Papers (Series I, Hogben, L.), University of Adelaide.

29. Fisher to Race, 27 September 1960, R. A. Fisher Papers (Series I, Race, R.R.), University of Adelaide.

30. I am indebted to Margaret Morrison for first bringing this anecdote to my attention.

31. Hogben, *LBL*, p. 213.

Chapter 3

1. Fuller and Thompson followed up *Behavior Genetics* (1960) with *Foundations of Behavior Genetics* (1978). By that time, though, Gerald E. McClearn and John C. DeFries published their *Introduction to Behavioral Genetics* (McClearn and DeFries 1973). McClearn and DeFries' text quickly became the standard introductory text for behavior genetics, undergoing half a dozen editions over four decades. The sixth edition was published in 2013 (Plomin, DeFries, Knopik, and Neiderhiser 2013).

2. This is the definition of heritability in the broad sense. This concept can be contrasted with heritability in the narrow sense, which only accounts for the propor-

tion of total phenotypic variance arising from the additive genetic component of genotypic variation. The IQ controversy mainly concerned heritability in the broad sense. Heritability in the narrow sense is of more interest to breeders.

3. For a probabilistic interpretation of heritability, see Tal (2009, 2012).

4. Jensen, to his credit, acknowledged this criticism very quickly. In response, he turned to other data sources (unblemished by charges of fabrication), which, he argued, continued to support the heritability of IQ at 0.8 (Jensen 1974).

5. For a selected bibliography of the IQ controversy, see Aby and McNamara (1990).

6. This chapter will deal with Lewontin's employment of genotype-environment interaction to attack Jensen. However, Lewontin also questioned Jensen's inference from within-group heritability to between-group heritability, as well as Jensen's treatment of variation within and between the races (Lewontin 1970a, 1972).

7. Dobzhansky, incidentally, was the first president of the Behavior Genetics Association. For his first presidential address, see Dobzhansky (1973).

8. Oddly, Lewontin makes no reference to Hogben in these discussions, even though their arguments were strikingly similar.

9. Feldman's association with Lewontin at this time had its own unique repercussions: shortly after co-authoring the *Science* article with Lewontin, Feldman wrote to his collaborator and explained that his recent job search was stymied because an administrator at the university which was considering hiring Feldman oversaw research on the genetics of alcoholism, schizophrenia, and criminality. Lewontin, though sorry for the inconvenience, seemed not at all surprised by the event: "Perhaps you will realize now that a close association with me has some real disadvantages, and that you might be wise to be a little more circumspect. I feel extremely bad about this and I urge you to consider the possibility that in the future you should be more cautious, especially where I am concerned." Feldman was ultimately offered the position after all. (Lewontin-Feldman correspondence, November-December 1976, Richard Lewontin Papers, Feldman File, American Philosophical Society [hereafter APS] Library, Accession Number B L59p.).

10. Lewontin to Dobzhansky, 7 January 1975, Theodosius Dobzhansky Papers (Lewontin file), APS Library, Accession Number B D65.

11. Lewontin to Dobzhansky, 2 May 1973, Dobzhansky Papers (Lewontin file), APS Library.

12. Sahotra Sarkar provides a very useful guide to reading Layzer's (1974) technical paper (Sarkar 1998).

13. Interaction is commonly abbreviated as "G×E."

14. Dobzhansky to Jensen, 17 January 1972, Dobzhansky Papers (Jensen file, folder #1), APS Library.

15. Jensen to Dobzhansky, 5 April 1974, Dobzhansky Papers (Jensen file, folder #1), APS Library.

16. Stephen Downes (2009) provides an insightful overview of the impact of Lewontin's arguments on the philosophy of biology.

17. For a sociological analysis of the divide, see Aaron Panofsky's "Field Analysis and Interdisciplinary Science: Scientific Capital Exchange in Behavior Genetics" (Panofsky 2011). Gry Oftedal offers a philosophical analysis of this debate (specifically between Lewontin and Sesardic); Oftedal points to Lewontin's focus on "structuring causes" with Sesardic's focus on "triggering causes" as the source of the divide, which overlaps with my focus on Lewontin's focus on causal mechanisms and Jensen's (and subsequently Sesardic's) focus on causes of variation (Oftedal 2005). I see Oftedal's contribution and mine in this chapter as of the same spirit, the difference being that I situate the dispute concerning causation within a broader debate about the distinct approaches to studying nature and nurture.

18. Similar points were made in these years between the eugenics controversy and the IQ controversy by Jane Loevinger (1943) and Anne Anastasi (1958). Loevinger pointed to interaction as an attack on assuming additivity in attempts to partition the relative contributions of nature and nurture to intelligence. Anastasi emphasized asking how questions rather than how-much questions.

19. For a similar point at this time, see Mather (1949, 49–51).

Chapter 4

1. The material for these next five paragraphs was drawn from personal interviews with Terrie Moffitt and Avshalom Caspi (18–19 February, 2013).

2. There has been a great deal of attention to this particular case of gene-environment interaction because of its association with criminal violence. Bioethicists have wondered (and, in some cases, even advocated for) genetic testing of embryos and newborns for their *MAOA* status. I discuss this issue in chapter 7.

3. Moffitt and Caspi's *5-HTTLPR*×stress study took second place to a paper published on the existence of dark matter and dark energy (Seife 2003).

4. Based on a Google Scholar citation count, available at http://scholar.google.com/scholar?q=caspi+moffitt&btnG=&hl=en&as_sdt=0%2C45.

5. For other philosophical discussions of this episode, see Schaffner (forthcoming, chapter 6) and Mitchell (2009, chapter 1). Schaffner discusses the episode in the context of recent developments in behavior genetics, which suggest the field is only

in its infancy. Mitchell introduces the case as an example of complexity in nature that does not submit to a reductionist approach to scientific explanation.

6. As will be discussed below, the exact count of independent replications depends on how you define "independent replication," and this was the subject of debate.

7. For other media reports on this meta-analysis, see Langreth (2009) in *Forbes*, Holden (2009) in *Science Now*, and Hamilton (2009) in *Time*.

8. For two different takes on the genome-wide association studies literature, see the favorable review by Visscher et al. (2012) and contrast it with the critical review by Turkheimer (2012).

9. Benedict Carey to Terrie Moffitt, 13 December, 2010.

10. This citation (Moffitt and Caspi 2009) is to a letter they prepared for reporters covering the Merikangas meta-analysis. They also wrote and submitted a reply to *JAMA*, where the Merikangas meta-analysis was published. But when the editors at *JAMA* edited their reply, Moffitt and Caspi chose not to publish the edited version. Douglas Kramer then published the content of Moffitt and Caspi's original letter to reporters in the American Academy of Child and Adolescent Psychiatry newsletter (Kramer 2009).

11. Interaction is often abbreviated as "G×E."

12. I am not claiming the explanatory divide that separated Fisher from Hogben and then Jensen from Lewontin explains everything about why the Merikangas and Munafò meta-analyses, on the one hand, and the Sen meta-analysis, on the other, came to different conclusions or why disputants might favor one meta-analysis over the other. For example, disputants on both sides of these meta-analyses have argued about data quantity versus data quality. Critics of the gene-environment interaction case involving the serotonin transporter gene and stress have pointed out that replications with the largest data sets more often do not replicate the original result (Munafò et al. 2010; Duncan and Keller 2011); defenders of the original result have pointed out that the large data sets are gathered with poor methods of measuring stress and depression (such as with self-reported questionnaires), while replications with smaller, high-quality data sets gathered with better methods of measuring stress and depression (such as with structured interviews) do replicate the original result (Uher and McGuffin 2008; Moffitt and Caspi 2009; Karg et al. 2011). The explanatory divide I discussed in the last chapters and bring to the present in this chapter has nothing to say about favoring data quantity or data quality.

Chapter 5

1. For a very helpful introduction to these four paradigms, see Schaffner (forthcoming, chapters 1 and 2).

2. Counterfactual dependence, for Woodward, is understood with the closely related concepts of intervention and invariance. An intervention consists of an idealized experimental manipulation of the value of some variable, thereby determining if it results in a change in the value of the outcome. So the counterfactuals are formulated in such a way that they show how the value of the outcome would change under the interventions that change the value of a variable; that is, they are formulated to show how the difference-makers make their difference. Invariance, then, is a characterization of the relationship between variables (or a variable and an outcome) under interventions on Woodward's account. When there is an invariant relationship between a variable and an outcome, then that relationship is potentially exploitable for manipulation, and because of this it is a causal relationship.

3. I thank Bill Bechtel for first pointing out to me this Platonic element.

4. There has been quite a bit of philosophical work done on Mayr's famous distinction between typological and population thinking (Sober 1980; Chung 2003; Ariew 2008; Godfrey-Smith 2009; Lewens 2009). Philosophers have asked, for example: in what sense did population thinking really begin with Darwin? And how did Mayr come to see the distinction? For our purposes here, however, the details of these debates are unnecessary. All I want to emphasize about the distinction is that many scientists ignore variation and focus on generalizations across individuals, while other scientists focus specifically on variation. As a result, my discussion of population thinking below should not be taken as representative of population thinking as a whole.

5. A notable exception to this is William Bechtel and Adele Abrahamsen's short discussion of variation in "Explanation: A Mechanist Alternative" (2005, 437–439). They note there that "variations are articulated in the description of the mechanism itself" (ibid., 439). My goal in this chapter can be thought of as an attempt to show how this articulation actually works.

6. This argument goes all the way back to John Stuart Mill (1843).

7. Note: the conditional "if" in this statement is ultimately an empirical matter. Philosophers who want to reply to Waters by claiming that DNA is not ontologically special cannot just point to potential difference makers. Rather, they must point to other difference makers that actually differed and were thus actual difference makers. So it is an empirical matter whether or not other molecules causally relevant to protein synthesis, such as RNA polymerase and ribosomes, are actual difference makers. For such a reply to Waters, see Stotz (2006).

8. Population thinking about mechanisms should not be confused with "mechanism thinking about populations"—the application of the philosophy of mechanisms to evolutionary questions about populations undergoing natural selection. For a debate over mechanism thinking about populations, see Skipper and Millstein

(2005), Barros (2008), Illari and Williamson (2010), Havstad (2011), Matthewson and Calcott (2011), and Matthews (2013). Brett Calcott (2009), however, does offer an amendment to the philosophy of mechanisms in the same spirit as that given here with his discussion of "lineage explanations." While my interest is in amending the philosophy of mechanisms to capture variation across populations, Calcott amends the philosophy of mechanisms to capture evolutionary changes that mechanisms undergo over time.

9. For the purposes of this discussion, I have only focused on BDNF's role in the activation of the NMDA receptor and the role of that molecular mechanism in the levels of mechanisms that are responsible for spatial memory. Scientists, however, have implicated BDNF in a variety of mechanisms and associated it with a variety of cognitive traits and disorders. For reviews, see Tyler et al. (2002), Bramham and Messaoudi (2005), and Blum and Konnerth (2005).

10. Daniel Weinberger, personal communication, 5 May 2012.

11. Note: because the Met/Met variant is so rare, researchers tend to just compare the Val/Val variant with the Val/Met variant, or group the Val/Met and the Met/Met variants together as a Met group.

12. For another application of this difference mechanism concept to the nature/nurture debate, see Karola Stotz's discussion of the transgenerational effect of maternal care behavior in rats (Stotz 2012).

13. On a related note, population thinking about mechanisms is not unique to the nature/nurture debate. Efforts to integrate variation-focused population thinking from one discipline with mechanism-elucidation from another discipline can be found across a range of interdisciplinary scientific efforts. Contemporary evolutionary developmental biology (or "evo-devo") seeks to unify population thinking from evolutionary biology with mechanism-elucidation from developmental biology (especially developmental genetics) in order to explain a phenomenon like morphological novelty (Brigandt and Love 2010). Likewise, anthropologists in the late 1940s and 1950s sought to create a "new physical anthropology" that brought traditional anthropology up to speed with the population thinking found in the burgeoning evolutionary synthesis (Smocovitis 2012). So my goal in this chapter is not to make the case for population thinking about mechanisms' uniqueness to the nature/nurture debate. On the contrary, I suspect it's quite common wherever population thinkers and mechanism-elucidators bump into one another because of a shared interest in a scientific problem. My goal in this chapter, rather, is to make sense of how population thinking about mechanisms works in the context of the nature/nurture debate. That model may or may not translate to the other domains.

14. Schaffner, in an earlier paper, referred to these as creeping or partial "reductions" (Schaffner 2006). However, in his most recent contribution, he clarifies:

The results are *like* reductions, but I think they are better described as *explanations*, using the term as an alternative to reduction because the e-word does not carry the conceptual freight of various reduction models and is a more appropriate general context within which to analyze what is actually occurring in the biomedical sciences. (author's emphasis, Schaffner forthcoming, chapter 5, 8).

Chapter 6

1. Refer back to figure 4.2 for the norm of reaction graph depicting this result.

2. Recommendations from a recent NIH-sponsored workshop on gene-environment interaction affirmed this integrative approach to studying interaction (Bookman et al. 2011).

3. Jonathan Kaplan (2000) provides a nice introduction to Cooper and Zubek's (1958) research and results.

4. Trevor Fuller, Sahotra Sarkar, and David Crews (2005) provide a useful survey of this type of norm of reaction research in rats and mice. Their review is insightful in showing how the original result of Cooper and Zubek (1958) continues to be borne out by more recent empirical studies.

5. Although less scrutinized, the serotonin transporter gene has been implicated in other cases of gene-environment interaction for traits such as inhibited behavior around unfamiliar peers (Fox et al. 2005), attention deficit hyperactivity disorder (Retz et al. 2008), and anxiety sensitivity (Stein, Schork, and Gelertner 2008).

Chapter 7

1. Readers who also read chapter 4 might wonder here about the robustness of Moffitt and Caspi's study of antisocial behavior, since the debate in chapter 4 was about the fact that different meta-analyses pointed in different directions with regard to the robustness of Moffitt and Caspi's study of depression (Caspi et al. 2003). Their *MAOA*×maltreatment study, however, seems to be holding up to scrutiny. Meta-analyses have affirmed the original interaction result (Kim-Cohen et al. 2006; Taylor and Kim-Cohen 2007; Byrd and Manuck forthcoming—in males but not females). Moreover, experimental research utilizing diverse methods ranging from neuroimaging to behavioral economics has supported the hypothesis that low-MAOA individuals are more prone to heightened emotional response then high-MAOA individuals when exposed to environmental stressors (Meyer-Lindberg et al. 2006; McDermott et al. 2009). That being said, I'm sure I speak for all of the bioethicists whom I discuss in this chapter in saying that subsequent research that questions the *MAOA*×maltreatment interaction effect would necessarily alter the bioethical reflections on the study.

2. For my technical definition of interaction, see the section in chapter 6 titled "The Conceptual Question." The more general definition above will suffice for the purposes of this chapter.

3. I should say that I think the mischaracterization I've diagnosed above is understandable. Media coverage of Moffitt and Caspi's *MAOA* study as well as the bioethical commentaries on it almost inevitably linked the results of their study to the earlier results of Brunner's *MAOA* study on the Dutch family. In the Dutch family, a genetically induced lack of MAOA was indeed the problem Brunner identified. So it was natural, when commentators transitioned to Moffitt and Caspi's study, to assume that less MAOA must again be the problem for the Dunedin population. The point, though, is that less MAOA is only the problem if you focus on one end of the environmental spectrum—the maltreated end.

4. Dobbs is finishing up a book-length treatment of this science, tentatively entitled *The Orchid and the Dandelion.*

5. In the United States, the most recent statistics from the U.S. Department of Health and Human Services Administration on Children, Youth and Families reported that approximately 681,000 children (or roughly 0.9%) were confirmed to be abused or neglected in 2011 (U.S. Department of Health and Human Services 2012, ix). As discussed earlier, though, what counts as evidence of maltreatment is much stricter for this population than for the Dunedin population, since these statistics result from actual child abuse investigations, while the Dunedin statistics result from investigations as well as parental reports of punishment habits and observer reports of parent-child interactions. So the percentage of children who are maltreated less severely or less conspicuously is certainly much higher than 0.9%. Just how much higher is open to speculation, but the likelihood that it is 50% or higher (which would tip the balance in favor of an intervention helping more than harming) seems small.

6. Campbell and Ross (2004) provided fascinating data from a survey of health care providers and parents about whether or not they would support and/or request genetic testing of children for violent predispositions. The health care providers in particular highlighted the dangers of a self-fulfilling prophecy; for example, one participant worried, "by telling people they're genetically predisposed, you might, subconsciously alter the way that they get treated and promote violence in them serendipitously" (ibid., 582).

7. There is another possibility that takes the differential susceptibility into consideration. Both the pharmaceutical and the social-behavioral interventions considered above involved screening all newborns and intervening on the low-MAOA children from birth; the problem with both interventions though, as I pointed out, is that it exposes the low-MAOA children who are not or would not be maltreated to all the costs of intervention without the benefits of it. But what about waiting until maltreatment has been established, then genetically screening and intervening on the

low-MAOA children? The idea here would be something like the MAOA booster shot for just the maltreated children or the social-behavioral intervention for just the maltreated children. By just focusing the intervention on the low-MAOA children who have already experienced maltreatment, my concern about harming the low-MAOA children who wouldn't have been maltreated is alleviated (Stone 2003; Lean 2012). This targeted intervention takes the reality of the differential susceptibility seriously; however, it faces its own problem. As developed in chapter 6, cases of gene-environment interaction reveal an interdependent causal relationship between the genetic variable and the environmental variable during the process of individual development. For antisocial behavior, this developmental story starts early, manifesting itself first as childhood conduct disorder and transitioning to antisocial personality disorder once the individual turns 18. During childhood, an individual learns the process of moral deliberation from those who care for him or her. This process of learning occurs at multiple levels: social, organismal, brain-system, cellular, and molecular. In emotionally charged situations, neurotransmitters are released throughout the brain; MAOA is then responsible for breaking down those neurotransmitters once the threat has passed. If caregivers maltreat the child through either neglect or abuse, these charged situations can become commonplace, and, without a sufficient supply of MAOA, heightened emotional response can persist for some time, ultimately having an impact on how the child learns to deal with conflict and resolution. Not surprisingly then, a premium is placed on early intervention in discussions of antisocial behavior treatment—focusing on children and adolescents (McCord and Tremblay 1992; Kim-Cohen et al. 2006). The rationale is that, once the environmental stressors are experienced and the disorder has developed, there is much less hope of effective treatment (Hemphill et al. 1998). So the problem with the targeted interventions envisioned above is that they have to wait until maltreatment has been confirmed before intervening. But, by that point, the devastating damage may already be done.

Chapter 8

1. See the previous chapter for a discussion of cases of interaction that change in rank and their characterization as constituting differential susceptibilities.

2. From the website of the National Institute of Allergy and Infectious Disease: www.niaid.nih.gov (content accessed 17 June, 2013).

3. For reviews of the hygiene hypothesis, see Schaub, Lauener, and von Mutius 2006 and von Mutius 2007.

4. For a replication of this interaction result in a much larger Dutch population, see Bottema et al. 2008.

5. Gern and Martinez, along with researchers at Wisconsin, Chicago, Arizona, and Roche Molecular Systems, utilized the COAST population to combine daycare expo-

sure with other genes associated with the immune system and also found interactions that change in rank for a variety of allergic responses (Hoffjan et al. 2005).

6. That same year (2008), Bakermans-Kranenburg and Van IJzendoorn published another study testing the differential effect of their parental sensitivity intervention on different variants of *DRD4*, but rather than tracking externalizing behavior as the behavioral measure they instead tracked cortisol levels in the children, which are an indicator of the children's hormonal stress level. They found the same result: 7+ repeat children who were exposed to the parental sensitivity intervention had lower basal cortisol levels than the 7+ children who did not; 7- children who were exposed to the parental sensitivity intervention, in contrast, had higher basal cortisol levels than the 7- children who did not (Bakermans-Kranenburg et al. 2008a).

7. For reviews of Bakermans-Kranenburg and van IJzendoorn's research, see Bakermans-Kranenburg and van IJzendoorn (2007, 2010).

8. In September 2013, the National Institutes of Health announced a $25 million research initiative to explore the promise of performing whole genome sequencing on newborns (Hayden 2013; Kaiser 2013).

9. The scenario for the pediatrician's office read as follows: "Imagine that you have a newborn baby. You take your child to the doctor for a routine check-up. During the visit, your child's doctor lets you know that you can get your child's whole genome sequenced. Testing would require a small sample of blood be taken. You would receive the results and would not have to pay for the testing. You can decide whether or not you want the information to be a part of your child's medical record. How interested would you be in getting your newborn baby's whole genome sequenced?"

10. See Tercyak et al. (2011) for a study of parental interest in predictive genetic testing of children through direct-to-consumer genetic testing; these authors also found parents to be interested in using the technology.

References

Aatre, Rajani D., and Sharlene M. Day. 2011. Psychological Issues in Genetic Testing for Inherited Cardiovascular Diseases. *Circulation: Cardiovascular Genetics* 4:81–90.

Aby, Stephen H., and Martha J. McNamara, eds. 1990. *The IQ Debate: A Selective Guide to the Literature*. New York: Greenwood Press.

Adams, Mark B. 1990. *The Wellborn Science: Eugenics in Germany, France, Brazil, and Russia*. Oxford: Oxford University Press.

Aguilera, M., B. Arias, M. Wichers, N. Barrantes-Vidal, J. Moya, H. Villa, J. van Os, et al. 2009. Early Adversity and 5-HTT/BDNF Genes: New Evidence of Gene-Environment Interactions on Depressive Symptoms in a General Population. *Psychological Medicine* 39:1425–1432.

Ainsworth, Mary D. Salter, Mary C. Blehar, Everett Waters, and Sally Wall. 1978. *Patterns of Attachment: A Psychological Study of the Strange Situation*. Hillsdale, NJ: Lawrence Erlbaum.

Allen, Garland E. 1986. The Eugenics Record Office, Cold Spring Harbor, 1910-1940: An Essay in Institutional History. *Osiris* 2:225–264.

Allen, Garland E. 1997. The Social and Economic Origins of Genetic Determinism: A Case History of the American Eugenics Movement, 1900–1940 and Its Lessons for Today. *Genetica* 99:77–88.

Allen, Garland E. 2011. Eugenics and Modern Biology: Critiques of Eugenics, 1910–1945. *Annals of Human Genetics* 75:314–325.

American Society of Human Genetics Board of Directors and the American College of Medical Genetics Board of Directors. 1995. Points to Consider: Ethical, Legal, and Psychological Implications of Genetic Testing in Children and Adolescents. *American Journal of Human Genetics* 57:1233–1241.

Anastasi, Anne. 1958. Heredity, Environment, and the Question 'How?' *Psychological Review* 65:197–208.

Andrews, Lori B., Jane E. Fullarton, Neil A. Holtzman, and Arno G. Motulsky, eds. 1994. *Assessing Genetic Risks: Implications for Health and Social Policy*. Washington, D.C.: National Academy Press.

Ankeny, Rachel A. 2000. Fashioning Descriptive Models in Biology: Of Worms and Wiring Diagrams. *Philosophy of Science* 67:S260–S272.

Araya, Ricardo, Xianzhang Hu, Jon Heron, Mary-Anne Enoch, Jonathan Evans, Glyn Lewis, David Nutt, et al. 2009. Effects of Stressful Life Events, Maternal Depression and 5-HTTLPR Genotype on Emotional Symptoms in Pre-Adolescent Children. *American Journal of Medical Genetics B: Neuropsychiatric Genetics* 150B:670–682.

Ariew, André. 2008. Population Thinking. In *Oxford Handbook of Philosophy of Biology*, ed. Michael Ruse, 64–86. Oxford University Press.

Baker, R. J. 1969. Genotype-Environment Interactions in Yield of Wheat. *Canadian Journal of Plant Science* 49:743–751.

Bakermans-Kranenburg, Marian J., and Marinus H. van IJzendoorn. 2006. Gene-Environment Interaction of the Dopamine D4 Receptor (DRD4) and Observed Maternal Insensitivity Predicting Externalizing Behavior in Preschoolers. *Developmental Psychobiology* 48:406–409.

Bakermans-Kranenburg, Marian J., and Marinus H. van IJzendoorn. 2007. Research Review: Genetic Vulnerability or Differential Susceptibility in Child Development: The Case of Attachment. *Journal of Child Psychology and Psychiatry and Allied Disciplines* 48:1160–1173.

Bakermans-Kranenburg, Marian J., and Marinus H. van IJzendoorn. 2010. Parenting Matters: Family Science in the Genomic Era. *Family Science* 1:26–36.

Bakermans-Kranenburg, Marian J., and Marinus H. van IJzendoorn. 2011. Differential Susceptibility to Rearing Environment Depending on Dopamine-Related Genes: New Evidence and a Meta-Analysis. *Development and Psychopathology* 23:39–52.

Bakermans-Kranenburg, Marian J., Marinus H. van IJzendoorn, Judi Mesman, Lenneke R. A. Alink, and Femmie Juffer. 2008a. Effects of an Attachment-Based Intervention on Daily Cortisol Moderated by Dopamine Receptor D4: A Randomized Control Trial on 1- to 3-Year Olds Screened for Externalizing Behavior. *Development and Psychopathology* 20:805–820.

Bakermans-Kranenburg, Marian J., Marinus H. van IJzendoorn, Femke T. A. Pijlman, Judi Mesman, and Femmie Juffer. 2008b. Experimental Evidence for Differential Susceptibility: Dopamine D4 Receptor Polymorphism (DRD4 VNTR) Moderates Intervention Effects on Toddlers' Externalizing Behavior in a Randomized Controlled Trial. *Developmental Psychology* 44:293–300.

Baldini, Mauro, I. Carla Lohman, Marilyn Halonen, Robert P. Erickson, Patrick G. Holt, and Fernando Martinez. 1999. A Polymorphism in the 5′ Flanking Region of the CD14 Gene Is Associated with Circulating Soluble CD14 Levels and with Total Serum Immunoglobulin E. *American Journal of Respiratory Cell and Molecular Biology* 20:976–983.

Barde, Yves-Alain. 1989. Trophic Factors and Neuronal Survival. *Neuron* 2: 1525–1534.

Barde, Yves-Alain, David Edgar, and Hans Thoenen. 1982. Purification of a New Neurotrophic Factor from Mammalian Brain. *EMBO Journal* 1:549–553.

Barkan, Elazar. 1991. Reevaluating Progressive Eugenics: Herbert Spencer Jennings and the 1924 Immigration Legislation. *Journal of the History of Biology* 24:91–112.

Barr, Christina S., Timothy K. Newman, Courtney Shannon, Clarissa Parker, Rachel L. Dvoskin, Michelle L. Becker, Melanie Schwandt, et al. 2004. Rearing Condition and rh5-HTTLPR Interact to Influence Limbic-Hypothalamic-Pituitary-Adrenal Axis Response to Stress in Infant Macaques. *Biological Psychiatry* 55:733–738.

Barros, D. Benjamin. 2008. Natural Selection as a Mechanism. *Philosophy of Science* 75:306–322.

Bartley, M. M. 1994. Conflicts in Human Progress: Sexual Selection and the Fisherian 'Runaway'. *British Journal for the History of Science* 27:177–196.

Bashford, Alison, and Philippa Levine. 2010. *The Oxford Handbook of the History of Eugenics*. Oxford: Oxford University Press.

Bath, Kevin G., and Francis S. Lee. 2006. Variant BDNF (Val66Met) Impact on Brain Structure and Function. *Cognitive, Affective & Behavioral Neuroscience* 6:79–85.

Bazelon, Emily. 2006. "A Question of Resilience." *The New York Times Magazine* (30 April). Available at http://www.nytimes.com/2006/04/30/magazine/30abuse.html?scp=1&sq=question%20of%20resilience&st=cse.

Bechtel, William. 2006. *Discovering Cell Mechanisms: The Creation of Modern Cell Biology*. Cambridge: Cambridge University Press.

Bechtel, William, and Adele Abrahamsen. 2005. Explanation: A Mechanist Alternative. *Studies in the History and Philosophy of Biological and Biomedical Sciences* 36:421–441.

Bechtel, William, and Robert C. Richardson. 1993. *Discovering Complexity: Decomposition and Localization as Strategies in Scientific Research*. Princeton: Princeton University Press.

Beevers, Christopher G., Brandon E. Gibb, John E. McGeary, and Ivan W. Miller. 2007. Serotonin Transporter Genetic Variation and Biased Attention for Emotional Word Stimuli among Psychiatric Inpatients. *Journal of Abnormal Psychology* 116:208–212.

Begley, Sharon. 2008. "But I Did Everything Right!" *The Daily Beast* (8 August). Available at http://www.newsweek.com/what-your-childs-dna-can-tell-you-about-parenting-87867.

Belsky, Jay. 1997. Variation in Susceptibility to Environmental Influence: An Evolutionary Argument. *Psychological Inquiry: An International Journal for the Advancement of Psychological Theory* 8:182–186.

Belsky, Jay, Marian J. Bakermans-Kranenburg, and Marinus H. van IJzendoorn. 2007. For Better *and* For Worse: Differential Susceptibility to Environmental Influences. *Current Directions in Psychological Science* 16:300–304.

Belsky, J., C. Jonassaint, M. Pluess, M. Stanton, B. Brummett, and R. Williams. 2009. Vulnerability Genes or Plasticity Genes? *Molecular Psychiatry* 14:746–754.

Belsky, Jay, and Michael Pluess. 2009. The Nature (and Nurture?) of Plasticity in Early Human Development. *Perspectives on Psychological Science* 4:345–351.

Benjamin, Jonathan, Lin Li, Chavis Patterson, Benjamin D. Greenberg, Dennis L. Murphy, and Dean H. Hamer. 1996. Population and Familial Association between the D4 Dopamine Receptor Gene and Measures of Novelty Seeking. *Nature Genetics* 12:81–84.

Bennett, J. H. 1983. *Natural Selection, Heredity, and Eugenics: Including Selected Correspondence of R.A. Fisher with Leonard Darwin and Others*. Oxford: Clarendon Press.

Bisgaard, Hans, Angela Simpson, Colin N. A. Palmer, Klaus Bønnelykke, Irwin Mclean, Somnath Mukhopadhyay, Christian B. Pipper, et al. 2008. Gene-Environment Interaction in the Onset of Eczema in Infancy: Filaggrin Loss-of-Function Mutations Enhanced by Neonatal Cat Exposure. PLoS Medicine 5: 0934–0940.

Blacker, C. P. 1952. *Eugenics: Galton and After*. Cambridge, MA: Harvard University Press.

Block, Ned. 1995. How Heritability Misleads About Race. *Cognition* 56:99–128.

Block, Ned, and Gerald Dworkin. 1976. IQ, Heritability and Inequality. In *The IQ Controversy*, ed. Ned Block and Gerald Dworkin, 410–540. New York: Pantheon.

Blot, William J., and Nicholas E. Day. 1979. Synergism and Interaction: Are They Equivalent? *American Journal of Epidemiology* 110:99–100.

Blum, Robert, and Arthur Konnerth. 2005. Neurotrophin-Mediated Rapid Signaling in the Central Nervous System: Mechanisms and Functions. *Physiology (Bethesda, MD)* 20:70–78.

Bookman, Ebony B., Kimberly McAllister, Elizabeth Gillanders, Kay Wanke, David Balshaw, Joni Rutter, Jill Reedy, et al. for the NIH G×E Interplay Workshop Participants. 2011. Gene-Environment Interplay in Common Complex Diseases: Forging an Integrative Model—Recommendations from an NIH Workshop. *Genetic Epidemiology* 35:217–225.

Borry, P. L. Stultiens, H. Nys, J.-J. Cassiman, and K. Dierickx. 2006. Presymptomatic and Predictive Genetic Testing in Minors: A Systematic Review of Guidelines and Position Papers. *Clinical Genetics* 70:374–381.

Bottema, R. W. B., N. E. Reijmerink, M. Kerkhof, G. H. Koppelman, F. F. Stelma, J. Gerritsen, C. Thijs, et al. 2008. *Interleukin 13, CD14,* Pet and Tobacco Smoke Influence Atopy in Three Dutch Cohorts: The Allergenic Study. *European Respiratory Journal* 32:593–602.

Bowler, Peter J. 2001. *Reconciling Science and Religion: The Debate in Early-Twentieth-Century Britain.* Chicago: University of Chicago Press.

Box, Joan Fisher. 1978. *R. A. Fisher: The Life of a Scientist.* New York: John Wiley and Sons.

Bramham, Clive R., and Elhoucine Messaoudi. 2005. BDNF Function in Adult Synaptic Plasticity: The Synaptic Consolidation Hypothesis. *Progress in Neurobiology* 76:99–125.

Brigandt, Ingo, and Alan C. Love. 2010. Evolutionary Novelty and the Evo-Devo Synthesis: Field Notes. *Evolutionary Biology* 37:93–99.

Briley, M. S., S. Z. Langer, R. Raisman, D. Sechter, and E. Zarifian. 1980. Tritiated Imipramine Binding Sites are Decreased in Platelets of Untreated Depressed Patients. *Science* 209:303–305.

Broberg, Gunnar, and Nils Roll-Hansen. 2005. *Eugenics and the Welfare State: Sterilization Policy in Denmark, Sweden, Norway, and Finland.* East Lansing, MI: Michigan State University Press.

Brooks-Crozier, Jennifer. 2011. The Nature and Nurture of Violence: Early Intervention Services for the Families of MAOA-Low Children as a Means to Reduce Violent Crime and the Costs of Violent Crime. *Connecticut Law Review* 44:531–573.

Brown, George W. 1987. Social Factors and the Development and Course of Depressive Disorder in Women: A Review of a Research Programme. *British Journal of Social Work* 17:615–634.

Brown, George W., Maria Ban, Thomas K. J. Craig, Tirril O. Harris, Joe Herbert, and Rudolf Uher. 2013. Serotonin Transporter Length Polymorphism, Childhood Maltreatment, and Chronic Depression: A Specific Gene-Environment Interaction. *Depression and Anxiety* 30:5–13.

Brummett, Beverly H., Stephen H. Boyle, Ilene C. Siegler, Cynthia M. Kuhn, Allison Ashley-Koch, Charles R. Jonassaint, Stephen Züchner, Ann Collins, and Redford B. Williams. 2008. Effects of Environmental Stress and Gender on Associations among

Symptoms of Depression and the Serotonin Transporter Gene Linked Polymorphic Region (5-HTTLPR). *Behavior Genetics* 38:34–43.

Brunner, Han G. 1996. MAOA Deficiency and Abnormal Behaviour: Perspectives on an Association. In *Genetics of Criminal and Antisocial Behaviour*, ed. Gregory R. Bock and Jamie A. Goode, 155–167. Chichester: John Wiley & Sons.

Brunner, H. G., M. Nelen, X. O. Breakefield, H. H. Ropers, and B. A. van Oost. 1993a. Abnormal Behavior Associated with a Point Mutation in the Structural Gene for Monoamine Oxidase A. *Science* 262:578–580.

Brunner, H. G., M. R. Nelen, P. van Zandvoort, N. G. G. M. Abeling, A. H. van Gennip, E. C. Wolters, M. A. Kuiper, et al. 1993b X-Linked Borderline Mental Retardation with Prominent Behavioral Disturbance: Phenotype, Genetic Localization, and Evidence for Disturbed Monoamine Metabolism. *American Journal of Human Genetics* 52:1032–1039.

Burbridge, David. 2001. Francis Galton on Twins, Heredity, and Social Class. *British Journal for the History of Science* 34:323–340.

Burt, Cyril. 1955. The Evidence for the Concept of Intelligence. *British Journal of Educational Psychology* 25:158–177.

Burt, Cyril. 1958. The Inheritance of Mental Ability. *American Psychologist* 13:1–15.

Burt, Cyril. 1966. The Genetic Determination of Differences in Intelligence: A Study of Monozygotic Twins Reared Together and Apart. *British Journal of Psychology* 57:137–153.

Byrd, Amy L., and Stephen B. Manuck. In press. *MAOA*, Childhood Maltreatment and Antisocial Behavior: Meta-Analysis of a Gene-Environment Interaction. *Biological Psychiatry*.

Cadoret, Remi J. 1982. Genotype-Environment Interaction in Antisocial Behaviour. *Psychological Medicine* 12:235–239.

Cadoret, Remi J., and Colleen Cain. 1981. Environmental and Genetic Factors in Predicting Adolescent Antisocial Behavior in Adoptees. *Psychiatric Journal of the University of Ottawa* 6:220–225.

Cadoret, Remi J., Colleen A. Cain, and Raymond R. Crowe. 1983. Evidence for Gene-Environment Interaction in the Development of Adolescent Antisocial Behavior. *Behavior Genetics* 13:301–310.

Cadoret, Remi J., William R. Yates, Ed Troughton, George Woodworth, and Mark A. Stewart. 1995. Genetic-Environmental Interaction in the Genesis of Aggressivity and Conduct Disorders. *Archives of General Psychiatry* 52:916–924.

Calcott, Brett. 2009. Lineage Explanations: Explaining How Biological Mechanisms Change. *British Journal for the Philosophy of Science* 60:51–78.

Campbell, E., and L. F. Ross. 2004. Attitudes of Healthcare Professionals and Parents Regarding Genetic Testing for Violent Traits in Childhood. *Journal of Medical Ethics* 30:580–586.

Canli, Turhan, and Klaus-Peter Lesch. 2007. Long Story Short: The Serotonin Transporter in Emotion Regulation and Social Cognition. *Nature Neuroscience* 10:1103–1109.

Carey, Benedict. 2009. "Report on Gene for Depression is Now Faulted." *The New York Times* (17 June). Available at http://www.nytimes.com/2009/06/17/science/17depress.html.

Carmichael, Leonard. 1925. Heredity and Environment: Are They Antithetical? *Journal of Abnormal and Social Psychology* 20:245–260.

Carola, Valeria, Giovanni Frazzetto, Tiziana Pascucci, Enrica Audero, Stefano Puglisi-Allegra, Simona Cabib, Klaus-Peter Lesch, and Cornelius Gross. 2008. Identifying Molecular Substrates in a Mouse Model of the Serotonin Transporter × Environment Risk Factor for Anxiety and Depression. *Biological Psychiatry* 63:840–846.

Caspi, Avshalom, Ahmad R. Hariri, Andrew Holmes, Rudolf Uher, and Terrie E. Moffitt. 2010. Genetic Sensitivity to the Environment: The Case of the Serotonin Transporter Gene and Its Implications for Studying Complex Diseases and Traits. *American Journal of Psychiatry* 167:509–527.

Caspi, Avshalom, Joseph McClay, Terrie E. Moffitt, Jonathan Mill, Judy Martin, Ian W. Craig, Alan Taylor, et al. 2002. Role of Genotype in the Cycle of Violence in Maltreated Children. *Science* 297:851–854.

Caspi, Avshalom, and Terrie E. Moffitt. 2006. Gene-Environment Interactions in Psychiatry: Joining Forces with Neuroscience. *Nature Reviews. Neuroscience* 7:583–590.

Caspi, Avshalom, Terrie E. Moffitt, Mary Cannon, Joseph McClay, Robin Murray, HonaLee Harrington, Alan Taylor, et al. 2005. Moderation of the Effect of Adolescent-Onset Cannabis Use on Adult Psychosis by a Functional Polymorphism in the Catechol-O-Methyltransferase Gene: Longitudinal Evidence of a Gene × Environment Interaction. *Biological Psychiatry* 57:1117–1127.

Caspi, Avshalom, Karen Sugden, Terrie E. Moffitt, Alan Taylor, Ian W. Craig, HonaLee Harrington, Joseph McClay, et al. 2003. Influence of Life Stress on Depression: Moderation by a Polymorphism in the 5-HTT Gene. *Science* 301:386–389.

Caspi, Avshalom, Benjamin Williams, Julia Kim-Cohen, Ian W. Craig, Barry J. Milne, Richie Poulton, Leonard C. Schalkwyk, Alan Taylor, Helen Werts, and Terrie E. Moffitt. 2007. Moderation of Breastfeeding Effects on the IQ by Genetic Variation in Fatty Acid Metabolism. *Proceedings of the National Academy of Sciences of the United States of America* 104:18860–18865.

Cervilla, J. A., E. Molina, M. Rivera, F. Torres-González, J. A. Bellón, B. Moreno, J. D. Luna, et al. 2007. The Risk for Depression Conferred by Stressful Life Events Is Modified by Variation at the Serotonin Transporter 5HTTLPR Genotype: Evidence from the Spanish PREDICT-Gene Cohort. *Molecular Psychiatry* 12: 748–755.

Champoux, M., A. Bennett, C. Shannon, J. D. Higley, K. P. Lesch, and S. J. Suomi. 2002. Serotonin Transporter Gene Polymorphism, Differential Early Rearing, and Behavior in Rhesus Monkey Neonates. *Molecular Psychiatry* 7:1058–1063.

Chase, Allan. 1980. *The Legacy of Malthus: The Social Costs of the New Scientific Racism.* Urbana: University of Illinois Press.

Chipman, P., A. F. Jorm, M. Prior, A. Sanson, D. Smart, X. Tan, and S. Easteal. 2007. No Interaction between the Serotonin Transporter Polymorphism (5-HTTLPR) and Childhood Adversity or Recent Stressful Life Events on Symptoms of Depression: Results from Two Community Surveys. *American Journal of Medical Genetics Part B: Neuropsychiatric Genetics* 144B:561–565.

Chorbov, Vesselin M., Elizabeth A. Lobos, Alexandre A. Todorov, Andrew C. Heath, Kelly N. Botteron, and Richard D. Todd. 2007. Relationship of 5-HTTLPR Genotypes and Depression Risk in the Presence of Trauma in a Female Twin Sample. *American Journal of Medical Genetics Part B: Neuropsychiatric Genetics* 144B:830–833.

Chung, Carl. 2003. On the Origin of the Typological/Population Distinction in Ernst Mayr's Changing Views of Species, 1942-1959. *Studies in the History and Philosophy of Biological and Biomedical Sciences* 34:277–296.

Clausen, Jens, David D. Keck, and William M. Hiesey. 1940. *Experimental Studies on the Nature of Species. I. Effect of Varied Environments on Western North American Plants.* Washington, D.C.: Carnegie Institution of Washington.

Cloninger, C. Robert, Sören Sigvardsson, Michael Bohman, and Anne-Liis von Knorring. 1982. Predisposition to Petty Criminality in Swedish Adoptees. *Archives of General Psychiatry* 39:1242–1247.

Cock, A. G. 1973. William Bateson, Mendelism, and Biometry. *Journal of the History of Biology* 6:1–36.

Collier, D. A., G. Stöber, T. Li, A. Heils, M. Catalano, D. Di Bella, M. J. Arranz, et al. 1996. A Novel Functional Polymorphism within the Promoter of the Serotonin Transporter Gene: Possible Role in Susceptibility to Affective Disorders. *Molecular Psychiatry* 1:453–460.

Colman, Andrew M., and Peter Woodhead. 1989. The Origin of the Juxtaposition of 'Nature' and 'Nurture': Not Galton, Shakespeare, or Mulcaster, but Socrates. *British Psychological Society History and Philosophy of Psychology Newsletter* 8:35–37.

Conley, James J. 1984. Not Galton, but Shakespeare: A Note on the Origin of the Term 'Nature and Nurture'. *Journal of the History of the Behavioral Sciences* 20:184–185.

Conrad, Peter. 2007. *The Medicalization of Society: On the Transformation of Human Conditions into Treatable Disorders*. Baltimore: Johns Hopkins University Press.

Conrad, Peter, Thomas Mackie, and Ateev Mehrotra. 2010. Estimating the Costs of Medicalization. *Social Science & Medicine* 70:1943–1947.

Cooper, R. M. and John P. Zubek. 1958. Effects of Enriched and Restricted Early Environments on the Learning Ability of Bright and Dull Rats. *Canadian Journal of Psychology* 12:159–164.

Covault, Jonathan, Howard Tennen, Stephen Armeli, Tamlin S. Conner, Aryeh I. Herman, Antonius H. N. Cillessen, and Henry R. Kranzler. 2007. Interactive Effects of the Serotonin Transporter 5-HTTLPR Polymorphism and Stressful Life Events on College Student Drinking and Drug Use. *Biological Psychiatry* 61:609–616.

Crain, Robert L. 1968. *The Politics of School Integration: Comparative Case Studies*. Chicago: Aldine Publishing Company.

Craver, Carl F. 2007. *Explaining the Brain: Mechanisms and the Mosaic Unity of Neuroscience*. Cambridge: Cambridge University Press.

Crews, F. A. E., W. J. Dakin, J. Heslop Harrison, Lancelot T. Hogben, Julian S. Huxley, J. Johnston, et al. 1923. The British Journal of Experimental Biology. *Science* 58:102.

Dahrendorf, Ralf. 1995. *LSE: A History of the London School of Economics and Political Science, 1895–1995*. Oxford: Oxford University Press.

Daniels, Norman. 1974. IQ, Heritability, and Human Nature. *PSA: Proceedings of the Biennial Meeting of the Philosophy of Science Association*, 143–180.

Dar-Nimrod, Ilan, and Steven J. Heine. 2011. Genetic Essentialism: On the Deceptive Determinism of DNA. *Psychological Bulletin* 137:800–818.

Darden, Lindley. 1977. William Bateson and the Promise of Mendelism. *Journal of the History of Biology* 10:87–106.

Darden, Lindley. 2006. *Reasoning in Biological Discoveries: Essays on Mechanisms, Interfield Relations, and Anomaly Resolution*. Cambridge: Cambridge University Press.

Darwin, Charles. 1859. *On the Origin of Species, by Means of Natural Selection or the Preservation of Favoured Races in the Struggle for Life*. London: John Murray.

Davison, Charlie. 1996. Predictive Genetics: The Cultural Implications of Supplying Probable Futures. In *The Troubled Helix: Social and Psychological Implications of the New Human Genetics*, ed. Theresa Marteau and Martin Richards, 317–330. Cambridge: Cambridge University Press.

DeWitt, Thomas J., and Samuel M. Scheiner, eds. 2004. *Phenotypic Plasticity: Functional and Conceptual Approaches*. Oxford: Oxford University Press.

Dierenfield, Bruce J. 2008. *The Civil Rights Movement. Revised Edition*. New York: Longman.

Dobbs, David. 2009. "The Science of Success." *The Atlantic* (9 December). Available at http://www.theatlantic.com/magazine/archive/2009/12/the-science-of-success/307761/.

Dobzhansky, Theodosius. 1973. Is Genetic Diversity Compatible with Human Equality? *Social Biology* 20:280–288.

Dobzhansky, Theodosius, and Boris Spassky. 1944. Genetics of Natural Populations. XXXIV. Adaptive Norm, Genetic Load, and Genetic Elite in *Drosophila pseudoobscura*. *Genetics* 48:1467–1485.

Downes, Stephen M. 2009. "Heritability." In *The Stanford Encyclopedia of Philosophy* (Fall 2010 Edition), ed. Edward N. Zalta. Available at http://plato.stanford.edu/archives/fall2010/entires/heredity/.

Dragunow, M. 1996. A Role for Immediate-Early Transcription Factors in Learning and Memory. *Behavior Genetics* 26:293–299.

Duncan, Laramie E., and Matthew C. Keller. 2011. A Critical Review of the First 10 Years of Candidate Gene-by-Environment Interaction Research in Psychiatry. *American Journal of Psychiatry* 168:1041–1049.

Dyson, Michael Eric. 2008. *April 4, 1968: Martin Luther King Jr.'s Death and the Transformation of America*. New York: Basic Books.

Eaves, Lindon J. 2006. Genotype × Environment Interaction in Psychopathology: Fact or Artifact? *Twin Research and Human Genetics* 9:1–8.

Ebstein, Richard P., Olga Novick, Roberto Umansky, Beatrice Priel, Yamima Osher, Darren Blaine, Estelle R. Bennett, Lubov Nemanov, Miri Katz, and Robert H. Belmaker. 1996. Dopamine D4 Receptor (*DRD4*) Exon III Polymorphism Associated with the Human Personality Trait of Novelty Seeking. *Nature Genetics* 12:78–80.

Eckberg, Douglas Lee. 1979. *Intelligence and Race: The Origins and Dimensions of the IQ Controversy*. New York: Praeger.

Economist, The. 2002. "The Origins of Violence: Nurturing Nature." (1 August). Available at http://www.economist.com/node/1259045.

Edwards, A. W. F. 1990. Fisher, W, and the Fundamental Theorem. *Theoretical Population Biology* 38:276–284.

Egan, Michael F., Masami Kojima, Joseph H. Callicott, Terry E. Goldberg, Bhaskar S. Kolachana, Alessandro Bertolino, Eugene Zaitsev, et al. 2003. The BDNF val66Met

Polymrophism Affects Activity-Dependent Secretion of BDNF and Human Memory and Hippocampal Function. *Cell* 112:257–269.

Eley, T. C., K. Sugden, A. Corsico, A. M. Gregory, P. Sham, P. McGuffin, R. Plomin, et al. 2004. Gene-Environment Interaction Analysis of Serotonin System Markers with Adolescent Depression. *Molecular Psychiatry* 9:908–915.

Ellis, Bruce J., W. Thomas Boyce, Jay Belsky, Marian J. Bakermans-Kranenburg, and Marinus H. van IJzendoorn. 2011. Differential Susceptibility to the Environment: An Evolutionary-Neurodevelopmental Theory. *Development and Psychopathology* 23:7–28.

Erlingsson, Steindor J. 2009. The Plymouth Laboratory and the Institutionalization of Experimental Zoology in Britain in the 1920s. *Journal of the History of Biology* 42:151–183.

Erspamer, V., and B. Asero. 1952. Identification of Enteramine, the Specific Hormone of the Enterochromaffin Cell System, as 5-Hydroxytryptamine. *Nature* 169: 800–801.

Eysenck, Hans J. 1971. *The IQ Argument: Race, Intelligence and Education.* LaSalle, Ill: Open Court Publishing.

Falconer, Douglas S. 1969. Quantitative Inheritance. In *Behavioral Genetics: Method and Research,* ed. Martin Manosevitz, Gardner Lindzey, and Delbert D. Thiessen, 120–142. New York: Appleton-Century-Crofts.

Feldman, Marcus W., and Richard C. Lewontin. 1975. The Heritability Hang-up. *Science* 190:1163–1168.

Fine, Sydney. 2000. *Expanding the Frontiers of Civil Rights: Michigan, 1948–1968.* Detroit: Wayne State University Press.

Fisher, Ronald A. 1918. The Correlation between Relatives on the Supposition of Mendelian Inheritance. *Transactions of the Royal Society of Edinburgh* 52:399–433.

Fisher, Ronald A. 1924a. The Elimination of Mental Defect. *Eugenics Review* 16:114–116.

Fisher, Ronald A. 1924b. The Biometrical Study of Heredity. *Eugenics Review* 16:189–210.

Fisher, Ronald A. 1925. *Statistical Methods for Research Workers.* Edinburgh: Oliver and Boyd Ltd.

Fisher, Ronald A. 1926a. Eugenics: Can It Solve the Problem of Decay of Civilizations? *Eugenics Review* 18:128–136.

Fisher, Ronald A. 1926b. Modern Eugenics, Being a Review of *The Need for Eugenic Reform,* by Leonard Darwin. *Eugenics Review* 18:231–236.

Fisher, Ronald A. 1932. Review of Lancelot Hogben's *Genetic Principles in Medicine and Social Science*. *Health and Empire* 7:147–150.

Fisher, Ronald A. 1951. *The Design of Experiments*. 6th ed. Edinburgh: Oliver and Boyd.

Fisher, Ronald A. 1955. Science and Christianity. *Friend* 113:42–43.

Fisher, Ronald A. 1958. *The Genetical Theory of Natural Selection*. New York: Dover.

Fisher, Ronald A., and Winifred A. Mackenzie. 1922. The Correlation of Weekly Rainfall. *Quarterly Journal of the Royal Meteorological Society* 48:234–245.

Fisher, Ronald A., and Winifred A. Mackenzie. 1923. Studies in Crop Variation. II. The Manurial Response of Different Potato Varieties. *Journal of Agricultural Science* 13:311–320.

Fisher, Ronald A., H. G. Thornton, and Winifred A. Mackenzie. 1922. The Accuracy of the Plating Method of Estimating the Density of Bacterial Populations. *Annals of Applied Biology* 9:325–359.

Fitzgerald, Deborah. 1990. *The Business of Breeding: Hybrid Corn in Illinois, 1890–1940*. Ithaca: Cornell University Press.

Flint, Jonathan, and Marcus Munafò. 2013. Candidate and Non-Candidate Genes in Behavior Genetics. *Current Opinion in Neurobiology* 23:57–61.

Foley, Debra L., Lindon J. Eaves, Brandon Wormley, Judy L. Silberg, Hermine H. Maes, Jonathan Kuhn, and Brien Riley. 2004. Childhood Adversity, Monoamine Oxidase A Genotype, and Risk for Conduct Disorder. *Archives of General Psychiatry* 61:738–744.

Forero, Diego A. 2009. "The Future of Genetic Studies of Complex Human Diseases": Drs. Merikangas and Risch Talk about That Influential Paper 13 Years Later. Hum-molgen.org (7 May). Available at http://hum-molgen.org/NewsGen/05-2009/000001.html.

Fox, Nathan A., Kate E. Nichols, Heather A. Henderson, Kenneth Rubin, Louis Schmidt, Dean Hamer, Monique Ernst, et al. 2005. Evidence for a Gene-Environment Interaction in Predicting Behavioral Inhibition in Middle Childhood. *Psychological Science* 16:921–926.

Fox Keller, Evelyn. 2000. *The Century of the Gene*. Cambridge: Harvard University Press.

Fox Keller, Evelyn. 2010. *The Mirage of a Space between Nature and Nurture*. Durham: Duke University Press.

Froggatt, P., and N. C. Nevin. 1971. The "Law of Ancestral Heredity" and the Mendelian-Ancestrian Controversy in England, 1889–1906. *Journal of Medical Genetics* 8:1–36.

Fuller, John L., and Edward C. Simmel. 1986. Trends in Behavior Genetics: 1960–1985. In *Perspectives in Behavior Genetics*, ed. John L. Fuller and Edward C. Simmel. Hillsdale, NJ: Lawrence Erlbaum.

Fuller, John L., and William R. Thompson. 1960. *Behavior Genetics*. New York: Wiley.

Fuller, John L., and William R. Thompson. 1978. *Foundations of Behavior Genetics*. Saint Louis: C.V. Mosby Company.

Fuller, Trevon, Sahotra Sarkar, and David Crews. 2005. The Use of Norms of Reaction to Analyze Genotypic and Environmental Influences on Behavior in Mice and Rats. *Neuroscience and Biobehavioral Reviews* 29:445–456.

Galton, Francis. 1865. Hereditary Talent and Character. *Macmillan's Magazine* 12:157–166, 318–327.

Galton, Francis. 1869. *Hereditary Genius, an Inquiry into Its Laws and Consequences*. London: Macmillan.

Galton, Francis. 1875. The History of Twins, as a Criterion of the Relative Powers of Nature and Nurture. *Fraser's Magazine* 12:566–576.

Galton, Francis. 1878. Composite Portraits. *Nature* 18:97–100.

Galton, Francis. 1904. Eugenics: Its Definition, Scope, and Aims. *American Journal of Sociology* 10:1–25.

Gelernter, J., A. J. Pakstis, and K. K. Kidd. 1995. Linkage Mapping of Serotonin Transporter Protein Gene SLC6A4 on Chromosome 17. *Human Genetics* 95:677–680.

Gern, James E., Claudia L. Reardon, Sabine Hoffjan, Dan Nicolae, Zhanhai Li, Kathy A. Roberg, William A. Neaville, et al. 2004. Effects of Dog Ownership and Genotype on Immune Development and Atopy in Infancy. *Journal of Allergy and Clinical Immunology* 113:307–314.

Gerstle, Gary. 2002. *American Crucible: Race and Nation in the Twentieth Century*. Princeton, NJ: Princeton University Press.

Gibbons, Ann. 2004. Tracking the Evolutionary History of the "Warrior" Gene. *Science* 304:818.

Gillespie, Nathan A., John B. Whitfield, Ben Williams, Andrew C. Heath, and Nicholas G. Martin. 2005. The Relationship between Stressful Life Events, the Serotonin Transporter (5-HTTLPR) Genotype and Major Depression. *Psychological Medicine* 35:101–111.

Gillette, Michael L. 2010. *Launching the War on Poverty: An Oral History*. 2nd ed. Oxford: Oxford University Press.

Gillham, Nicholas Wright. 2001. *A Life of Sir Francis Galton: From African Exploration to the Birth of Eugenics*. Oxford: Oxford University Press.

Glennan, Stuart. 1996. Mechanisms and the Nature of Causation. *Erkenntnis* 44:49–71.

Glennan, Stuart. 2002. Rethinking Mechanistic Explanation. *Philosophy of Science* 69:S342–S353.

Godfrey-Smith, Peter. 2009. *Darwinian Populations and Natural Selection*. Oxford: Oxford University Press.

Goldenberg, Aaron J., Daniel S. Dodson, Matthew M. Davis, and Beth A. Tarini. 2014. Parents' Interest in Whole-Genome Sequencing of Newborns. *Genetics in Medicine* 16:78–84.

Gottesman, Irving I., and James Shields. 1967. A Polygenic Theory of Schizophrenia. *Proceedings of the National Academy of Sciences of the United States of America* 58: 199–205.

Gottlieb, Gilbert. 1992. *Individual Development and Evolution: The Genesis of Novel Behavior*. New York: Oxford University Press.

Gottlieb, Gilbert. 1995. Some Conceptual Deficiencies in "Developmental" Behavior Genetics. *Human Development* 38:131–141.

Gottlieb, Gilbert. 2003. On Making Behavior Genetics Truly Developmental. *Human Development* 46:337–355.

Grabe, H. J., M. Lange, B. Wolff, H. Völzke, M. Lucht, H. J. Freyberger, U. John, et al. 2005. Mental and Physical Distress Is Modulated by a Polymorphism in the 5-HT Transporter Gene Interacting with Social Stressors and Chronic Disease Burden. *Molecular Psychiatry* 10:220–224.

Grafius, J. E. 1956. The Interaction of Genotype and Night Temperature in Oat and Barley Varieties. *Agronomy Journal* 48:56–59.

Griffiths, Paul E. 2006. The Fearless Vampire Conservator: Philip Kitcher, Genetic Determinism, and the Informational Gene. In *Genes in Development: Re-Reading the Molecular Paradigm*, ed. Eva M. Neumann-Held and Christoph Rehmann-Sutter, 175–198. Durham: Duke University Press.

Griffiths, Paul E., and James Tabery. 2013. Developmental Systems Theory: What Does It Explain, and How Does It Explain It? *Advances in Child Development and Behavior* 44:65–94.

Gurdon, John B., and Nick Hopwood. 2000. The Introduction of *Xenopus laevis* into Developmental Biology: Of Empire, Pregnancy Testing and Ribosomal Genes. *International Journal of Developmental Biology* 44:43–50.

Haberstick, Brett C., Jeffrey M. Lessem, Christian J. Hopfer, Andrew Smolen, Marissa A. Ehringer, David Timberlake, and John K. Hewitt. 2005. Monoamine Oxidsase A (MAOA) and Antisocial Behaviors in the Presence of Childhood and Adolescent

Maltreatment. *American Journal of Medical Genetics Part B: Neuropsychiatric Genetics* 135B:59–64.

Hacking, Ian. 1994. Styles of Scientific Thinking or Reasoning: A New Analytical Tool for Historians and Philosophers of the Sciences. In *Trends in the Historiography of Science*, ed. Kostas Gavrolugu, Jean Christianidis, and Efthymios Nicolaidis, 31–48. Dordrecht, Netherlands: Kluwer Academic.

Hagen, Joel. 2003. The Statistical Frame of Mind in Systematic Biology from *Quantitative Zoology* to *Biometry*. *Journal of the History of Biology* 36:353–384.

Haldane, J. B. S. 1932. A Programme for Human Genetics: A Review of Lancelot Hogben's *Genetic Principles in Medicine and Social Science*. *Nature* 129:345–346.

Haldane, J. B. S. 1936. Some Principles of Causal Analysis in Genetics. *Erkenntnis* 6:346–357.

Haldane, J. B. S. 1946. The Interaction of Nature and Nurture. *Annals of Eugenics* 13:197–205.

Hamer, Dean. 2002. Rethinking Behavior Genetics. *Science* 298:71–72.

Hamilton, Anita. 2009. "Study: 'Depression Gene' Doesn't Predict the Blues." *Time* (17 June). Available at http://www.time.com/time/health/article/0,8599,1905083,00.html.

Harden, K. Paige, Eric Turkheimer, and John C. Loehlin. 2006. Genotype by Environment Interaction in Adolescents' Cognitive Aptitude. *Behavior Genetics* 37:273–283.

Hardy, John, and Nancy C. Low. 2011. Genes and Environment in Psychiatry: Winner's Curse or Cure? *Archives of General Psychiatry* 68:455–456.

Hariri, Ahmad R., Emily M. Drabant, Karen E. Munoz, Bhaskar S. Kolachana, Venkata S. Mattay, Michael F. Egan, and Daniel R. Weinberger. 2005. A Susceptibility Gene for Affective Disorders and the Response of the Human Amygdala. *Archives of General Psychiatry* 62:146–152.

Hariri, Ahmad R., Terry E. Goldberg, Venkata S. Mattay, Bhaskar S. Kolachana, Joseph H. Callicott, Michael F. Egan, and Daniel R. Weinberger. 2003. Brain-Derived Neurotrophic Factor val66met Polymorphism Affects Human Memory-Related Hippocampal Activity and Predicts Memory Performance. *Journal of Neuroscience* 23:6690–6694.

Hariri, Ahmad R., Venkata S. Mattay, Alessandro Tessitore, Bhaskar Kolachana, Francesco Fera, David Goldman, Michael F. Egan, et al. 2002. Serotonin Transporter Genetic Variation and the Response of the Human Amygdala. *Science* 297: 400–403.

Harris-Zsovan, Jane. 2010. *Eugenics and the Firewall: Canada's Nasty Little Secret*. Winnipeg: J. Gordon Shillingford Publishing.

Havstad, Joyce C. 2011. Discussion: Problems for Natural Selection as a Mechanism. *Philosophy of Science* 78:512–523.

Hayden, Erika Check. 2009. Genome Sequencing: The Third Generation. *Nature* 457:768–769.

Hayden, Erika Check. 2013. Scientists to Sequence Genomes of Hundreds of Newborns. *Nature* News Blog. Available online http://blogs.nature.com/news/2013/09/scientists-to-sequence-hundreds-of-newborns-genomes.html.

Heils, A., A. Teufel, S. Petri, M. Seemann, D. Bengel, U. Balling, P. Riederer, et al. 1995. Functional Promoter and Polyadenylation Site Mapping of the Human Serotonin (5-HT) Transporter Gene. *Journal of Neural Transmission* 102:247–254.

Heils, Armin, Andreas Teufel, Susanne Petri, Gerald Stöber, Peter Riederer, Dietmar Bengel, and K. Peter Lesch. 1996. Allelic Variation of Human Serotonin Transporter Gene Expression. *Journal of Neurochemistry* 66:2621–2624.

Hempel, Carl G., and Paul Oppenheim. 1948. Studies in the Logic of Explanation. *Philosophy of Science* 15:135–175.

Hemphill, James F., Ron Templeman, Stephen Wong, and Robert D. Hare. 1998. Psychopathy and Crime: Recidivism and Criminal Careers. In *Psychopathy: Theory, Research, and Implications for Society*, ed. David J. Cooke, Adelle E. Forth, and Robert D. D. Hare, 375–398. Dordrecht: Kluwer.

Herrnstein, Richard J. 1971. I.Q. *Atlantic Monthly* 228:43–64.

Hodge, M. J. S. 1992. Biology and Philosophy (Including Ideology): A Study of Fisher and Wright. In *The Founders of Evolutionary Genetics: A Centenary Reappraisal*, ed. Sahotra Sarkar, 231–293. Dordrecht: Kluwer.

Hoffjan, Sabine, Dan Nicolae, Irina Ostrovnaya, Kathy Roberg, Michael Evans, Daniel B. Mirel, Lori Steiner, et al. 2005. Gene-Environment Interaction Effects on the Development of Immune Responses in the 1st Year of Life. *American Journal of Human Genetics* 76:696–704.

Hoffmann, Diane E., and Eric A. Wulfsberg. 1995. Testing Children for Genetic Predispositions: Is It in Their Best Interest? *Journal of Law, Medicine & Ethics* 23:331–344.

Hogben, Lancelot. 1927. *Principles of Evolutionary Biology*. Cape Town: Juta.

Hogben, Lancelot. 1931a. *The Nature of Living Matter*. New York: Alfred A. Knopf.

Hogben, Lancelot. 1931b. The Foundations of Social Biology. *Economica* 31:4–24.

Hogben, Lancelot. 1932. *Genetic Principles in Medicine and Social Science*. New York: Alfred A. Knopf.

Hogben, Lancelot. 1933a. *Nature and Nurture, Being the William Withering Memorial Lectures*. London: George Allen and Unwin (originally published by Williams & Norgate).

Hogben, Lancelot. 1933b. The Limits of Applicability of Correlation Technique in Human Genetics. *Journal of Genetics* 27:379–406.

Hogben, Lancelot. 1937. *Mathematics for the Million*. New York: W. W. Norton.

Hogben, Lancelot. 1938. *Science for the Citizen: A Self-Educator Based on the Social Background of Scientific Discovery*. New York: Alfred A. Knopf.

Hogben, Lancelot. 1998. *Lancelot Hogben, Scientific Humanist: An Unauthorized Autobiography*. Suffolk: Merlin Press.

Hogben, Lancelot T., and Myer Sim. 1953. The Self-Controlled and Self-Recorded Clinical Trial for Low-Grade Morbidity. [Reprinted in 2011, *International Journal of Epidemiology* 40:1438–1454.] *British Journal of Preventive & Social Medicine* 7:163–179.

Hogben, Lancelot T., and Frank R. Winton. 1922a. The Pigmentary Effector System. I.—Reaction of Frog's Melanophores to Pituitary Extracts. *Proceedings of the Royal Society of London: Series B, Containing Papers of a Biological Character* 93:318–329.

Hogben, Lancelot T., and Frank R. Winton. 1922b. The Pigmentary Effector System. II. *Proceedings of the Royal Society of London: Series B, Containing Papers of a Biological Character* 94:151–162.

Hogben, Lancelot T., and Frank R. Winton. 1923. The Pigmentary Effector System. III.—Colour Response in the Hypophysectomised Frog. *Proceedings of the Royal Society of London: Series B, Containing Papers of a Biological Character* 95:15–31.

Holden, Constance. 2009. "Sad News for 'Depression Gene'." *ScienceNOW* (16 June). Available at http://news.sciencemag.org/sciencenow/2009/06/16-01.html.

Hu, Xian-Zhang, Robert H. Lipsky, Guanshan Zhu, Longina A. Akhtar, Julie Taubman, Benjamin D. Greenberg, Ke Xu, et al. 2006. Serotonin Transporter Promoter Gain-of-Function Genotypes are Linked to Obsessive-Compulsive Disorder. *American Journal of Human Genetics* 78:815–826.

Hunt, J. M. 1969. Has Compensatory Education Failed? Has It Been Attempted? *Harvard Educational Review* 39:278–300.

Huxley, Julian. 1932. Eugenics: A Review of Lancelot Hogben's *Genetic Principles in Medicine and Social Science. Eugenics Review* 23:341–344.

Huxley, Julian. 1942. *Evolution. The Modern Synthesis*. London: Allen and Unwin.

Huxley, Julian S., and Lancelot T. Hogben. 1922. Experiments on Amphibian Metamorphosis and Pigment Responses in Relation to Internal Secretions. *Proceedings of the Royal Society of London: Series B, Containing Papers of a Biological Character* 93:36–53.

Illari, Phyllis McKay, and Jon Williamson. 2010. Function and Organization: Comparing the Mechanisms of Protein Synthesis and Natural Selection. *Studies in the History and Philosophy of Biological and Biomedical Sciences* 41:279–291.

Ioannidis, John P. A. 2005. Why Most Published Research Findings Are False. *PLoS Medicine* 2:e124.

Jackson, John P. 2005. *Science for Segregation: Race, Law, and the Case Against Brown v. Board of Education.* New York: New York University Press.

Jacobs, Nele, Gunter Kenis, Frenk Peeters, Catherine Derom, Robert Vlietinck, and Jim van Os. 2006. Stress-Related Negative Affectivity and Genetically Altered Serotonin Transporter Function: Evidence of Syngergism in Shaping Risk of Depression. *Archives of General Psychiatry* 63:989–996.

Jacobs, Patricia A., Muriel Brunton, Marie M. Melville, R. P. Brittain, and W. F. McClemont. 1965. Aggressive Behaviour, Mental Sub-normality, and the *XYY* Male. *Nature* 208:1351–1352.

Jamieson, Annie and Gregory Radick. 2013. Putting Mendel in His Place: How Curriculum Reform in Genetics and Counterfactual History of Science Can Work Together. In *The Philosophy of Biology: A Companion for Educators*, ed. Kostas Kampourakis, 577–596. Dordrecht: Springer.

Jensen, Arthur R. 1969. How Much Can We Boost IQ and Scholastic Achievement? *Harvard Educational Review* 39:1–123.

Jensen, Arthur R. 1970. Race and the Genetics of Intelligence: A Reply to Lewontin. *Bulletin of the Atomic Scientists* 26:17–23.

Jensen, Arthur R. 1972. The IQ Controversy: A Reply to Layzer. *Cognition* 1:427–452.

Jensen, Arthur R. 1973. *Educability and Group Differences.* New York: Harper and Row.

Jensen, Arthur R. 1974. Kinship Correlations Reported by Sir Cyril Burt. *Behavior Genetics* 4:1–28.

Jensen, Arthur R. 1975. The Meaning of Heritability in the Behavioral Sciences. *Educational Psychologist* 11:171–183.

Jensen, Arthur R. 1976. Heritability of IQ. *Science* 194:6–8.

Jinks, J. L., and David W. Fulker. 1970. Comparison of the Biometrical Genetical, MAVA, and Classical Approaches to the Analysis of Human Behavior. *Psychological Bulletin* 73:311–349.

Johnstone, D. J. 1987. Tests of Significance Following R.A. Fisher. *British Journal for the Philosophy of Science* 38:481–499.

Jones, Kevin R., and Louis F. Reichardt. 1990. Molecular Cloning of a Human Gene That Is a Member of the Nerve Growth Factor Family. *Proceedings of the National Academy of Sciences of the United States of America* 87:8060–8064.

Jonsson, Per. 1959. Investigations of Group Feeding versus Individual Feeding and on the Interaction between Genotype and Environment in Pigs. *Acta Agricultura Scandinavica* 9:204–228.

Kaiser, Jocelyn. 2013. NIH Studies Explore Promise of Sequencing Babies' Genomes. *Science* Insider. Available at http://news.sciencemag.org/biology/2013/09/nih-studies-explore-promise-sequencing-babies%E2%80%99-genomes.

Kallmann, Franz J. 1954. Twin Data in the Analysis of Mechanisms of Inheritance. *American Journal of Human Genetics* 6:157–174.

Kamin, Leon J. 1974. *The Science and Politics of I.Q.* Potomac, MD: Lawrence Erlbaum.

Kang, H., and E. M. Schuman. 1995. Long-Lasting Neurotrophin-Induced Enhancement of Synaptic Transmission in the Adult Hippocampus. *Science* 267:1658–1662.

Kanner, Baruch I., and Shimon Schuldiner. 1987. Mechanism of Transport and Storage of Neurotransmitters. *CRC Critical Reviews in Biochemistry* 22:1–38.

Kaplan, Jonathan Michael. 2000. *The Limits and Lies of Human Genetic Research*. New York: Routledge.

Kaplan, Karen. 2007. "IQ Isn't Fed by Genes Alone, a Study on Breast Milk Finds." *Los Angeles Times* (10 November). Available at http://www.latimes.com/features/health/la-na-stemcell111007,0,3204431.story.

Karg, Katja, Margit Burmeister, Kerby Shedden, and Srijan Sen. 2011. The Serotonin Transporter Promoter Variant (5-HTTLPR, Stress, and Depression Meta-Analysis Revisited. *Archives of General Psychiatry* 68:444–454.

Karnik, Meghana S., Lei Wang, Deanna M. Barch, John C. Morris, and John G. Csernanssky. 2010. BDNF Polymorphism rs6265 and Hippocampal Structure and Memory Performance in Healthy Control Subjects. *Psychiatry Research* 178:425–429.

Katsuyama, Hironobu, Masafumi Tomita, Kazuo Hidaka, Shigeko Fushimi, Toshiko Okuyama, Yoko Watanabe, Yoshie Tamechika, et al. 2008. Association between Serotonin Transporter Gene Polymorphism and Depressed Mood Caused by Job Stress in Japanese Workers. *International Journal of Molecular Medicine* 21:499–505.

Kaufman, Joan, Joel Gelertner, Arie Kaffman, Avshalom Caspi, and Terrie Moffitt. 2010. Arguable Assumptions, Debatable Conclusions. *Biological Psychiatry* 67:e19–e20.

Kaufman, Joan, Bao-Zhu Yang, Heather Douglas-Palumberi, Damion Grasso, Deborah Lipschitz, Shadi Houshyar, John H. Krystal, et al. 2006. Brain-Derived Neurotrophic Factor—5-HTTLPR Gene Interactions and Environmental Modifiers of Depression in Children. *Biological Psychiatry* 59:673–680.

Kaufman, Joan, Bao-Zhu Yang, Heather Douglas-Palumberi, Shadi Houshyar, Deborah Lipschitz, John H. Krystal, and Joel Gelernter. 2004. Social Supports and Serotonin Transporter Gene Moderate Depression in Maltreated Children. *Proceedings of the National Academy of Sciences of the United States of America* 101:17316–17321.

Kempthorne, Oscar. 1974. A Review of Collected Papers of R. A. Fisher (Vol. 1), ed. J. H. Bennett. *Social Biology* 21:98–101.

Kendler, Kenneth S. 2005. Psychiatric Genetics: A Methodologic Critique. *American Journal of Psychiatry* 162:3–11.

Kendler, Kenneth S., Jonathan W. Kuhn, Jen Vittum, Carol A. Prescott, and Brien Riley. 2005. The Interaction of Stressful Life Events and a Serotonin Transporter Polymorphism in the Prediction of Episodes of Major Depression: A Replication. *Archives of General Psychiatry* 62:529–535.

Kevles, Daniel J. 1995. *In the Name of Eugenics: Genetics and the Uses of Human Heredity*. Cambridge, Mass: Harvard University Press.

Kilpatrick, Dean G., Karestan C. Koenen, Kenneth J. Ruggiero, Ron Acierno, Sandro Galea, Heidi S. Resnick, John Roitzsch, et al. 2007. The Serotonin Transporter Genotype and Social Support and Moderation of Posttraumatic Stress Disorder and Depression in Hurricane-Exposed Adults. *American Journal of Psychiatry* 164:1693–1699.

Kim, Jae-Min, Robert Stewart, Sung-Wan Kim, Su-Jin Yang, Il-Seon Shin, Young-Hoon Kim, and Jin-Sang Yoon. 2007. Interactions between Life Stressors and Susceptibility Genes (5-HTTLPR and BDNF) on Depression in Korean Elders. *Biological Psychiatry* 62:423–428.

Kim-Cohen, J., A. Caspi, A. Taylor, B. Williams, R. Newcombe, I. W. Craig, and T. E. Moffitt. 2006. *MAOA*, Maltreatment, and Gene-Environment Interaction Predicting Children's Mental Health: New Evidence and a Meta-Analysis. *Molecular Psychiatry* 11:903–913.

King, J. W. B. 1963. A Genotype-Environment Interaction Experiment with Bacon Pigs. *Animal Production* 5:283–288.

King, J. W. B., J. H. Watson, and G. B. Young. 1959. Genotype-Environment Interactions in the Wintering of Lambs. *Journal of Agricultural Science* 53:156–171.

King, J. W. B., and G. B. Young. 1955. A Study of Three Breeds of Sheep Wintered in Four Environments. *Journal of Agricultural Science* 45:331–338.

Kitcher, Philip. 2001. Battling the Undead: How (and How Not) to Resist Genetic Determinism. In *Thinking About Evolution: Historical, Philosophical, and Political Perspectives*, ed. Rama S. Singh, Costas B. Krimbas, Diane B. Paul, and John Beatty, 396–414. Cambridge: Cambridge University Press.

Krafka, J. 1920. The Effect of Temperature upon Facet Number in the Bar-Eyed Mutant of Drosophila. *Journal of General Physiology* 2:409–464.

Kramer, Douglas A. 2009. G×E Depression Hypothesis Challenged: Researchers Reply. *AACAP News* November/December: 283–284.

Kristjansson, F. K. 1957. Observations on Genotype-Environment Interaction in Swine. *Canadian Journal of Animal Science* 37:179–184.

Krout, Maurice H. 1931. Heredity, Environment, and Developmental Process. *Psychological Review* 38:187–211.

Kühl, Stefan. 1994. *The Nazi Connection: Eugenics, American Racism, and German National Socialism*. Oxford: Oxford University Press.

Kuhn, Thomas S. 1962. *The Structure of Scientific Revolutions*. Chicago: University of Chicago Press.

Kumar, Ravenish. 2003. Violence Begets Violence; or Does It? A Brain Enzyme Protects Victimes of Childhood Abuse from Becoming Antisocial and Criminal. *Clinical Genetics* 63:180–183.

Kupper, Lawrence L., and Michael D. Hogan. 1978. Interaction in Epidemiologic Studies. *American Journal of Epidemiology* 108:447–453.

Lakatos, Imre. 1977. *The Methodology of Scientific Research Programmes: Philosophical Papers*, vol. 1. Cambridge: Cambridge University Press.

Lander, Eric S., and Nicholas J. Schork. 1994. Genetic Dissection of Complex Traits. *Science* 265:2037–2048.

Laney, Garrine P. 2003. *The Voting Rights Act of 1965: Historical Background and Current Issues*. Hauppauge: Novinka Books.

Langer, S. Z., E. Zarifian, M. Briley, R. Raisman, and D. Sechter. 1981. High-Affinity Binding of 3H-Imipramine in Brain and Platelets and Its Relevance to the Biochemistry of Affective Disorders. *Life Sciences* 29:211–220.

Langreth, Robert. 2009. Hunt for Depression Gene Goes Into Reverse. *Forbes.com* (16 June). Available at http://www.forbes.com/2009/06/16/depression-genetics-jama-business-healthcare-genes.html.

Largent, Mark A. 2007. *Breeding Contempt: The History of Coerced Sterilization in the United States*. New Brunswick: Rutgers University Press.

Laucht, Manfred, Jens Treutlein, Dorothea Blomeyer, Arlette F. Buchmann, Brigitte Schmid, Katja Becker, Ulrich S. Zimmermann, et al. 2009. Interaction between the 5-HTTLPR Serotonin Transporter Polymorphism and Environmental Adversity for Mood and Anxiety Psychopathology: Evidence from a High-Risk Community Sample of Young Adults. *International Journal of Neuropsychopharmacology* 12:737–747.

Laudan, Larry. 1977. *Progress and Its Problems.* Berkeley: University of California Press.

Layzer, David. 1972. Science or Superstition? (A Physical Scientist Looks at the IQ Controversy). *Cognition* 1:265–299, 453–473.

Layzer, David. 1974. Heritability Analyses of IQ Scores: Science or Numerology? *Science* 183:1259–1266.

Lazary, Judit, Aron Lazary, Xenia Gonda, Anita Benko, Eszter Molnar, Gabriella Juhasz, and Gyorgy Bagdy. 2008. New Evidence for the Association of the Serotonin Transporter Gene (SLCC6A4) Haplotypes, Threatening Life Events, and Depressive Phenotype. *Biological Psychiatry* 64:498–504.

Lean, Christopher. 2012. Reducing Reactive Aggression: MAOA and Responsible Intervention. Presented at the 64th Annual Northwest Philosophy Conference, 27 October. Corvallis: Oregon State University.

Leibrock, Joachim, Friedrich Lottspeich, Andreas Hohn, Magdalena Hofer, Bastian Hengerer, Piotr Masiakowski, Hans Thoenen, et al. 1989. Molecular Cloning and Expression of Brain-Derived Neurotrophic Factor. *Nature* 341:149–152.

Lenze, Eric J., Michael C. Munin, Robert E. Ferrell, Bruce G. Pollock, Elizabeth Skidmore, Francis Lotrich, Joan C. Rogers, et al. 2005. Association of the Serotonin Transporter Gene-Linked Polymorphic Region (5-HTTLPR) Genotype with Depression in Elderly Persons after Hip Surgery. *American Journal of Geriatric Psychiatry* 13:428–432.

Lerner, Isadore Michael. 1950. *Population Genetics and Animal Improvement.* Cambridge: Cambridge University Press.

Lesch, K.-P., U. Balling, J. Gross, K. Strauss, B. L. Wolozin, D. L. Murphy, and P. Riederer. 1994. Organization of the Human Serotonin Transporter Gene. *Journal of Neural Transmission* 95:157–162.

Lesch, Klaus-Peter, Dietmar Bengel, Armin Heils, Sue Z. Sabol, Benjamin D. Greenberg, Susanne Petri, Jonathan Benjamin, et al. 1996. Association of Anxiety-Related Traits with a Polymorphism in the Serotonin Transporter Gene Regulatory Region. *Science* 274:1527–1531.

Levin, Michael. 1997. *Why Race Matters: Race Differences and What They Mean.* Westport, CT: Praeger.

Levine, Eric S., Robert A. Crozier, Ira B. Black, and Mark R. Plummer. 1998. Brain-Derived Neurotrophic Factor Modulates Hippocampal Synaptic Transmission by Increasing N-Methyl-D-Aspartic Acid Receptor Activity. *Proceedings of the National Academy of Sciences of the United States of America* 95:10235–10239.

Levine, Eric S., Cheryl F. Dreyfus, Ira B. Black, and Mark R. Plummer. 1995. Brain-Derived Neurotrophic Factor Rapidly Enhances Synaptic Transmission in Hippocampal Neurons Via Postsynaptic Tyrosine Kinase Receptors. *Proceedings of the National Academy of Sciences of the United States of America* 92:8074–8077.

Lewens, Tim. 2009. What Is Wrong with Typological Thinking? *Philosophy of Science* 76:355–371.

Lewontin, Richard C. 1970a. Race and Intelligence. *Bulletin of the Atomic Scientists* 26:2–8.

Lewontin, Richard C. 1970b. Further Remarks on Race and the Genetics of Intelligence. *Bulletin of the Atomic Scientists* 26:23–25.

Lewontin, Richard C. 1972. The Apportionment of Human Diversity. *Evolutionary Biology* 6:381–398.

Lewontin, Richard C. 1974. The Analysis of Variance and the Analysis of Causes. *American Journal of Human Genetics* 26:400–411.

Lewontin, Richard C. 1975. Genetic Aspects of Intelligence. *Annual Review of Genetics* 9:387–405.

Lewontin, Richard C. 1976. Sociobiology as an Adaptationist Paradigm. *Behavioral Science* 24:5–14.

Lewontin, Richard C., Diane Paul, John Beatty, and Costas B. Krimbas. 2001. Interview of R. C. Lewontin. In *Thinking about Evolution: Historical, Philosophical, and Political Perspectives*, ed. Rama S. Singh, Costas B. Krimbas, Diane B. Paul, and John Beatty, 22–61. Cambridge: Cambridge University Press.

Lickliter, Robert and Cheryl Logan. 2007. Developmental Psychobiology Special Issue: "Gilbert Gottlieb's Legacy: Probabilistic Epigenesis and the Development of Individuals and Species". *Developmental Psychobiology* 49:747–748.

Lickliter, Robert and Hunter Honeycutt. In press. Biology, Development, and Human Systems. In *Handbook of Child Psychology and Developmental Science, Volume 1: Theory & Method (Seventh Edition)*, ed. Willis F. Overton, Peter C. M. Molenaar, and Richard M. Lerner. Wiley Blackwell.

Lindzey, Gardner, and Delbert D. Thiessen, eds. 1970. *Contributions to Behavior-Genetic Analysis: The Mouse as Prototype*. New York: Appleton-Century-Crofts.

Liungman, Carl. 1974. *What is IQ?: Intelligence, Heredity, and Environment*. London: Gordon and Cremonesi.

Loehlin, John C. 1989. Partitioning Environmental and Genetic Contributions to Behavioral Development. *American Psychologist* 44:1285–1292.

Loevinger, Jane. 1943. On the Proportional Contributions of Differences in Nature and in Nurture to Differences in Intelligence. *Psychological Bulletin* 40:725–756.

Loevy, Robert D. 1997. *The Civil Rights Act of 1964: The Passage of the Law That Ended Racial Segregation*. Albany: State University of New York.

Lombardo, Paul A. 1996. Medicine, Eugenics, and the Supreme Court: From Coercive Sterilization to Reproductive Freedom. *Journal of Contemporary Health Law and Policy* 13:1–25.

Lombroso, Cesare. 1876. *L'Uomo Delinquente*. Milan: Hoepli.

London, Stephanie J., and Isabelle Romieu. 2009. Gene by Environment Interaction in Asthma. *Annual Review of Public Health* 30:55–80.

Longino, Helen. 2001. What Do We Measure When We Measure Aggression? *Studies in History and Philosophy of Science, Part A* 32:685–704.

Longino, Helen. 2013. *Studying Human Behavior: How Scientists Investigate Aggression and Sexuality*. Chicago: University of Chicago Press.

Ludmerer, Kenneth M. 1972. *Genetics and American Society: A Historical Appraisal*. Baltimore: Johns Hopkins University Press.

Lumpkins, Charles L. 2008. *American Pogrom: The East St. Louis Race Riot and Black Politics*. Athens: Ohio University Press.

Lush, Jay L. 1937. *Animal Breeding Plans*. Ames: Collegiate Press, Inc.

Lush, Jay L. 1951. The Impact of Genetics on Animal Breeding. *Journal of Animal Science* 10:311–321.

Lynch, Michael, and Bruce Walsh. 1997. *Genetics and Analysis of Quantitative Traits*. Sunderland, MA: Sinauer Associates.

Lyons, Richard D. 1968a. "Genetic Abnormality Is Linked to Crime; Genetics Linked to Violent Crimes." *The New York Times*. 21 April, page 1.

Lyons, Richard D. 1968b. "Ultimate Speck Appeal May Cite a Genetic Defect." *The New York Times*. 22 April, page 43.

Lyons, Richard D. 1968c. "Chromosome Test for Flaws Costly; 2 Scientists Cite Scarcity of Skilled Aides for Analyses." *The New York Times*. 23 April, page 27.

Machamer, Peter K., Lindley Darden, and Carl F. Craver. 2000. Thinking about Mechanisms. *Philosophy of Science* 67:1–25.

MacKenzie, Donald A. 1981a. Sociobiologists in Competition: The Biometry-Mendelism Controversy. In *Biology, Medicine, and Society 1840–1940*, ed. C. Webster, 243–288. Cambridge: Cambridge University Press.

MacKenzie, Donald A. 1981b. *Statistics in Britain: 1865–1930, The Social Construction of Scientific Knowledge*. Edinburgh: Edinburgh University Press.

Macones, George A., Samuel Parry, Mohammed Elkousy, Bonnie Clothier, Serdar H. Ural, and Jerome F. Strauss. 2004. A Polymorphism in the Promoter Region of TNF and Bacterial Vaginosis: Preliminary Evidence of Gene-Environment Interaction in the Etiology of Spontaneous Preterm Birth. *American Journal of Obstetrics and Gynecology* 190:1504–1508.

Mahalanobis, P. C. 1964. Professor Ronald Aylmer Fisher. *Biometrics* 20:238–250.

Maisonpierre, Peter C., Michelle M. Le Beau, Rafael Espinosa, III, Nancy Y. Ip, Leonardo Belluscio, Suzanne M. de la Monte, Stephen Squinto, et al. 1991. Human and Rat Brain-Derived Neurotrophic Factor and Neurotrophin-3: Gene Structures, Distributions, and Chromosomal Localizations. *Genomics* 10:558–568.

Mandelli, Laura, Alessandro Serretti, Elena Marino, Adele Pirovano, Raffaella Calati, and Cristina Colombo. 2007. Interaction between Serotonin Transporter Gene, Catechol-O-Methyltransferase Gene and Stressful Life Events in Mood Disorders. *International Journal of Neuropsychopharmacology* 10:437–447.

Mann, Denise. 2011. "'Depression Gene' Linked to Response in Stress." *WebMD* (4 January). Available at http://www.webmd.com/depression/news/20110104/depresssion-gene-linked-to-response-to-stress.

Marcus, Pamela M., Richard B. Hayes, Paolo Vineis, Montserrat Garcia-Closas, Neil E. Caporaso, Herman Autrup, Robert A. Branch, et al. 2000. Cigarette Smoking, N-Acetyltransferase 2 Acetylation Status, and Bladder Cancer Risk: A Case-Series Meta-Analysis of a Gene-Environment Interaction. *Cancer Epidemiology, Biomarkers & Prevention* 9:461–467.

Martinez, Fernando D. 2005. Gene-Environment Interactions in Asthma and Allergies: A New Paradigm to Understand Disease Causation. *Immunology and Allergy Clinics of North America* 25:709–721.

Mather, Kenneth. 1949. *Biometrical Genetics: The Study of Continuous Variation*. New York: Dover Publications.

Mather, Kenneth. 1964. R.A. Fisher's Work in Genetics. *Biometrics* 20:330–342.

Matsuo, Keitaro, Nobuyuki Hamajima, Masayuki Shinoda, Shunzo Hatooka, Manami Inoue, Toshiro Takezaki, and Kazuo Tajima. 2001. Gene-Environment Interaction between an Aldehyde dehydrogenase-2 (ALDH2) Polymorphism and Alcohol Consumption for the Risk of Esophageal Cancer. *Carcinogenesis* 22:913–916.

Matthews, Lucas. 2013. Conflicting Results for Natural Selection and the New Philosophy of Mechanism. International Society for the History, Philosophy, and Social Studies of Biology, 11 July, 2013, Montpellier, France.

Matthews, Michael R. 1980. *The Marxist Theory of Schooling: A Study of Epistemology and Education*. Sussex: Harvester Press.

Matthewson, John, and Brett Calcott. 2011. Mechanistic Models of Population-Level Phenomena. *Biology and Philosophy* 26:737–756.

Mayr, Ernst. 1975. Typological versus Population Thinking. In *Evolution and the Diversity of Life*, ed. Ernst Mayr, 26–29. Cambridge: Harvard University Press.

Mazumdar, Pauline M. H. 1992. *Eugenics, Human Genetics, and Human Failings: The Eugenics Society, Its Sources and Its Critics in Britain*. London: Routledge.

McClearn, Gerald E., and John C. DeFries. 1973. *Introduction to Behavioral Genetics*. San Francisco: W. H. Freeman.

McCord, Joan, and Richard E. Tremblay, eds. 1992. *Preventing Antisocial Behavior: Interventions from Birth through Adolescence*. New York: Guilford Press.

McDermott, Rose, Dustin Tingley, Jonathan Cowden, Giovanni Frazzetto, and Dominic D. P. Johnson. 2009. Monoamine Oxidase A Gene (MAOA) Predicts Behavioral Aggression Following Provocation. *Proceedings of the National Academy of Sciences of the United States of America* 106:2118–2123.

McGue, M., and T. J. Bouchard, Jr. 1989. Genetic and Environmental Determinants of Informational Processing and Special Mental Abilities. In *Advances in the Psychology of Human Intelligence*, vol. 5, ed. R. J. Sternberg, 7–45. Hillsdale, NJ: Lawrence Erlbaum.

Mednick, Sarnoff A., William F. Gabrielli, and Barry Hutchings. 1984. Genetic Influences in Criminal Convictions: Evidence from an Adoption Cohort. *Science* 224:891–894.

Meltzer, Herbert Y. 1990. Role of Serotonin in Depression. *Annals of the New York Academy of Sciences* 600:486–500.

Meltzer, H. Y., and R. C. Arora. 1988. Genetic Control of Serotonin Uptake in Blood Platelets: A Twin Study. *Psychiatry Research* 24:263–269.

Meredith, William. 1968. Factor Analysis and the Use of Inbred Strains. In *Progress in Human Behavior Genetics: Recent Reports on Genetic Syndromes, Twin Studies, and Statistical Advances*, ed. Steven G. Vandenberg, 335–348. Baltimore: Johns Hopkins University Press.

Merikangas, Kathleen Ries, Thomas Lehner, and Neil J. Risch. 2009. In Reply. *Journal of the American Medical Association* 302:1861–1862.

Mestel, Rosie. 2011. "Depression Gene: Its Rise, Its Fall, Its Rise?" *Los Angeles Times* (3 January). Available at http://articles.latimes.com/print/2011/jan/03/news/lat-depression-gene-its-rise-its-f-01032011.

Meyer-Lindenberg, Andreas, Joshua W. Buckholtz, Bhaskar Kolachana, Ahmad R. Hariri, Lukas Pezawas, Giuseppe Blasi, Ashley Wabnitz, et al. 2006. Neural Mechanisms of Genetic Risk for Impulsivity and Violence in Humans. *Proceedings of the National Academy of Sciences of the United States of America* 103:6269–6274.

Middeldorp, C. M., D. C. Cath, A. L. Beem, G. Willemsen, and D. I. Boomsma. 2008. Life Events, Anxious Depression and Personality: A Prospective and Genetic Study. *Psychological Medicine* 38:1557–1565.

Middeldorp, Christel M., Eco J. C. de Geus, A. Leo Beem, Nico Lakenberg, Jouke-Jan Hottenga, P. Eline Slagboom, and Dorret I. Boomsma. 2007. Family Based Association Analyses between the Serotonin Transporter Gene Polymorphism (5-HTTLPR) and Neuroticism, Anxiety, and Depression. *Behavior Genetics* 37:294–301.

Miele, Frank. 2002. *Intelligence, Race, and Genetics: Conversations with Arthur R. Jensen.* Boulder: Westview.

Mill, John Stuart. 1843. *A System of Logic, Ractiocinative, and Inductive: Being a Connected View of the Principles of Evidence and the Methods of Scientific Investigation.* London: John W. Parker.

Mill, J. S., A. Caspi, J. McClay, K. Sugden, S. Purcell, P. Asherson, I. Craig, et al. 2002. The Dopamine D4 Receptor and the Hyperactivity Phenotype: A Developmental-Epidemiological Study. *Molecular Psychiatry* 7:383–391.

Mitchell, Sandra D. 2003. *Biological Complexity and Integrative Pluralism.* Cambridge: Cambridge University Press.

Mitchell, Sandra D. 2009. *Unsimple Truths: Science, Complexity, and Policy.* Chicago: University of Chicago Press.

Mizuno, Makoto, Kiyofumi Yamada, Jue He, Akira Nakajima, and Toshitaka Nabeshima. 2003. Involvement of BDNF Receptor TrkB in Spatial Memory Formation. *Learning & Memory (Cold Spring Harbor, NY)* 10:108–115.

Mizuno, Makoto, Kiyofumi Yamada, Ana Olariu, Hiroyuki Nawa, and Toshitaka Nabeshima. 2000. Involvement of Brain-Derived Neurotrophic Factor in Spatial Memory Formation and Maintenance in a Radial Arm Maze Test in Rats. *Journal of Neuroscience* 20:7116–7121.

Moffitt, Terrie E., and Avshalom Caspi. 2009. "Drs. Moffitt and Caspi's Response Letter for the Press." Available at http://www.moffittcaspi.com/GxE_Serotonin.

Moffitt, Terrie E., Avshalom Caspi, and Michael Rutter. 2005. Strategy for Investigating Interactions between Measured Genes and Measured Environments. *Archives of General Psychiatry* 62:473–481.

Moffitt, Terrie E., Avshalom Caspi, Michael Rutter, and Phil A. Silva. 2001. *Sex Differences in Antisocial Behaviour: Conduct Disorder, Delinquency, and Violence in the Dunedin Longitudinal Study*. Cambridge: Cambridge University Press.

Moore, David S. 2002. *The Dependent Gene: The Fallacy of "Nature vs. Nurture."*. New York: Henry Holt/Times Books.

Moore, David S. 2006. A Very Little Bit of Knowledge: Re-Evaluating the Meaning of the Heritability of IQ. *Human Development* 49:347–353.

Moore, David S. 2008a. Individuals and Populations: How Biology's Theory and Data Have Interfered with the Integration of Development and Evolution. *New Ideas in Psychology* 26:370–386.

Moore, David S. 2008b. Espousing Interactions and Fielding Reactions: Addressing Laypeople's Beliefs about Genetic Determinism. *Philosophical Psychology* 21:331–348.

Moore, James. 2007. R. A. Fisher: A Faith Fit for Eugenics. *Studies in the History and Philosophy of Biological and Biomedical Sciences* 38:110–135.

Moran, Mark. 2006. Jury Still Out on Impact of Genes on Trial Verdicts. *Psychiatric News* 41:12.

Moran, P. A. P., and C. A. B. Smith. 1966. Commentary on R. A. Fisher's Paper on the Correlation Between Relatives on the Supposition of Mendelian Inheritance. *Eugenics Laboratory Memoirs* 41:1–62.

Morell, Virginia. 1993. Evidence Found for a Possible "Aggression Gene". *Science* 260:1722–1723.

Moreno, Jonathan D. 2003. Neuroethics: An Agenda for Neuroscience and Society. *Nature Reviews: Neuroscience* 4:149–153.

Morley, F. H. W. 1956. Selection for Economic Characters in Australian Merino Sheep. *Australian Journal of Agricultural Research* 7:140–146.

Morley, F. H. W., and C. I. Davern. 1956. Flowering Time in Subterranean Clover. *Australian Journal of Agricultural Research* 7:388–400.

Morley, Katherine I., and Grant W. Montgomery. 2001. The Genetics of Cognitive Processes: Candidate Genes in Humans and Animals. *Behavior Genetics* 31:511–531.

Munafò, Marcus R., Caroline Durrant, Glyn Lewis, and Jonathan Flint. 2009. Gene × Environment Interactions at the Serotonin Transporter Locus. *Biological Psychiatry* 65:211–219.

Munafò, Marcus R., Caroline Durrant, Glyn Lewis, and Jonathan Flint. 2010. Defining Replication: A Response to Kaufman and Colleagues. *Biological Psychiatry* 67:e21–e23.

Murray, Jay C., and Laval M. Verhalan. 1970. Genotype by Environment Interaction Study of Cotton in Oklahoma. *Crop Science* 10:197–199.

Nakatani, Daisaku, Hiroshi Sato, Yasuhiko Sakata, Issei Shiotani, Kunihiro Kinjo, Hiroya Mizuno, Masahiko Shimizu, et al., and the Osaka Acute Coronary Insufficiency Study (OACIS) Group. 2005. Influence of Serotonin Transporter Gene Polymorphism on Depressive Symptoms and New Cardiac Events after Acute Myocardial Infarction. *American Heart Journal* 150:652–658.

Nelson, Nicole C. 2013. Modeling Mouse, Human, and Discipline: Epistemic Scaffolds in Animal Behavior Genetics. *Social Studies of Science* 43:3–29.

News and Editorial Staffs. 2003. The Runners-Up. *Science* 302:2039–2045.

Nichols, Robert C. 1969. The Resemblance of Twins in Personality and Interests. In *Behavioral Genetics: Method and Research*, ed. Martin Manosevitz, Gardner Lindzey, and Delbert D. Thiessen, 580–596. New York: Appleton-Century-Crofts.

Nichols, Robert C. 1978. Twin Studies of Ability, Personality, and Interests. *Homo* 29:158–173.

Nilsson, Kent W., Rickard L. Sjöberg, Mattias Damberg, Jerzy Leppert, John Öhrvik, Per Olof Alm, Leif Lindström, et al. 2006. Role of Monoamine Oxidase A Genotype and Psychosocial Factors in Male Adolescent Criminal Activity. *Biological Psychiatry* 59:121–127.

Nisbett, Richard E., Joshua Aronson, Clancy Blair, William Dickens, James Flynn, Diane F. Halpern, and Eric Turkheimer. 2012. Intelligence: New Findings and Theoretical Developments. *American Psychologist* 67:130–159.

Norton, Bernard J. 1973. The Biometric Defense of Darwinism. *Journal of the History of Biology* 6:283–316.

Norton, Bernard J. 1975. Biology and Philosophy: The Methodological Foundations of Biometry. *Journal of the History of Biology* 8:85–93.

Norton, Bernard J. 1978. Fisher and the Neo-Darwinian Synthesis. In *Human Implications of Scientific Advance: Proceedings of the XVth International Congress of the History of Science*, ed. E. G. Forbes, 481–494. Edinburgh: Edinburgh University Press.

Norton, Bernard, and E. S. Pearson. 1976. A Note on the Background to, and Refereeing of, R. A. Fisher's 1918 Paper "On the Correlation between Relatives on the Supposition of Mendelian Inheritance". *Notes and Records of the Royal Society of London* 31:151–162.

Ober, Carole. 2005. Perspectives on the Past Decade of Asthma Genetics. *Journal of Allergy and Clinical Immunology* 116:274–278.

Ober, Carole, and Sabine Hoffjan. 2006. Asthma Genetics 2006: The Long and Winding Road to Gene Discovery. *Genes and Immunity* 7:95–100.

Office of National Drug Control Policy. 2007. *The Link between Marijuana & Mental Illness: A Survey of Recent Research*. Washington, DC: Executive Office of the President of the United States.

Oftedal, Gry. 2005. Heritability and Genetic Causation. *Philosophy of Science* 72:699–709.

Olby, Robert. 1988. The Dimensions of Scientific Controversy: The Biometric-Mendelian Debate. *British Journal for the History of Science* 22:299–320.

Olszynko-Gryn, Jesse. 2013. When Pregnancy Tests Were Toads: The *Xenopus* Test in the Early NHS. *Wellcome History* 51:2–3.

Ordover, Nancy. 2003. *American Eugenics: Race, Queer Anatomy, and the Science of Nationalism*. Minneapolis: University of Minnesota Press.

Osborn, Frederick. 1974. History of the American Eugenics Society. *Biodemography and Social Biology* 21:115–126.

Oyama, Susan. 1985. *The Ontogeny of Information: Developmental Systems and Evolution*. Cambridge: Cambridge University Press.

Oyama, Susan. 2000. Causal Democracy and Causal Contributions in Developmental Systems Theory. *Philosophy of Science* 67:S332–S347.

Panofsky, Aaron L. 2011 Field Analysis and Interdisciplinary Science: Scientific Capital Exchange in Behavior Genetics. *Minerva* 49:295–316.

Parens, Erik. 2004. Genetic Differences and Human Identities: On Why Talking about Behavioral Genetics Is Important and Difficult. *Hastings Center Report* 34 (Suppl.):S1–S36.

Parens, Erik. 2013. On Good and Bad Forms of Medicalization. *Bioethics* 27:28–35.

Patterson, Susan L., Ted Abel, Thomas A. S. Deuel, Kelsey C. Martin, Jack C. Rose, and Eric R. Kandel. 1996. Recombinant BDNF Rescues Deficits in Basal Synaptic Transmission and Hippocampal LTP in BDNF Knockout Mice. *Neuron* 16: 1137–1145.

Paul, Diane B. 1995. *Controlling Human Heredity: 1865 to the Present*. Atlantic Highlands, NJ: Humanities Press.

Paul, Diane B. 1998. *The Politics of Heredity: Essays on Eugenics, Biomedicine, and the Nature-Nurture Debate*. Albany: State University of New York Press.

Paykel, Eugene S. 1978. Contribution of Life Events to Causation of Psychiatric Illness. *Psychological Medicine* 8:245–253.

Pearson, Karl. 1914. *Birth 1822 to Marriage 1853*, vol. 1. The Life, Letters, and Labours of Francis Galton. Cambridge: Cambridge University Press.

Pearson, Karl. 1924. *Researches in Middle Life*, vol. 2. The Life, Letters, and Labours of Francis Galton. Cambridge: Cambridge University Press.

Pearson, Karl. 1930a. *The Life, Letters, and Labours of Francis Galton, vol. 3A: Correlation, Personal Identification and Eugenics*. Cambridge: Cambridge University Press.

Pearson, Karl. 1930b. *The Life, Letters, and Labours of Francis Galton, vol. 3B: Characterization, especially by Letters, & Index*. Cambridge: Cambridge University Press.

Perbal, Laurence. 2013. G×E Interaction and Pluralism in the Postgenomic Era. *Biological Theory* 7:266–274.

Pernick, Martin S. 1996. *The Black Stork: Eugenics and the Death of "Defective" Babies in American Medicine and Motion Pictures Since 1915*. Oxford: Oxford University Press.

Pezawas, Lukas, Andreas Meyer-Lindenberg, Emily M. Drabant, Beth A. Verchinski, Karen E. Munoz, Bhaskar S. Kolachana, Michael F. Egan, et al. 2005. 5-HTTLPR Polymorphism Impacts Human Cingulate-Amygdala Interactions: A Genetic Susceptibility Mechanism for Depression. *Nature Neuroscience* 8:828–834.

Pezawas, Lukas, Beth A. Verchinski, Venkata S. Mattay, Joseph H. Callicott, Bhaskar S. Kolachana, Richard E. Straub, Michael F. Egan, et al. 2004. The Brain-Derived Neurotrophic Factor val66met Polymorphism and Variation in Human Cortical Morphology. *Journal of Neuroscience* 24:10099–10102.

Pigliucci, Massimo. 2001. *Phenotypic Plasticity: Beyond Nature and Nurture*. Baltimore: Johns Hopkins University Press.

Piotrowska, Monika. 2012. From Humanized Mice to Human Disease: Guiding Extrapolation from Model to Target. *Biology and Philosophy* 28:439–455.

Plomin, Robert, and John C. DeFries. 1976. Heritability of IQ. *Science* 194:10–12.

Plomin, Robert, and John C. DeFries. 1983. The Colorado Adoption Project. *Child Development* 54:276–289.

Plomin, Robert, and John C. DeFries. 1985. *Origins of Individual Differences in Infancy: The Colorado Adoption Project*. New York: Academic Press.

Plomin, Robert, John C. DeFries, and David W. Fulker. 1988. *Nature and Nurture during Infancy and Early Childhood*. New York: Cambridge University Press.

Plomin, Robert, John C. DeFries, Valerie S. Knopik, and Jenae M. Neiderhiser. 2013. *Behavioral Genetics*. 6th ed. New York: Worth Publishers.

Plomin, Robert, John C. DeFries, and John C. Loehlin. 1977. Genotype-Environment Interaction and Correlation in the Analysis of Human Behavior. *Psychological Bulletin* 84:309–322.

Plomin, Robert, J. C. DeFries, and G. E. McClearn. 1990. *Behavioral Genetics: A Primer*. 2nd ed. New York: W. H. Freeman.

Plomin, Robert, Michael J. Owen, and Peter McGuffin. 1994. The Genetic Basis of Complex Human Behaviors. *Science* 264: 1733–1739.

Plutynski, Anya. 2005. Parsimony and the Fisher-Wright Debate. *Biology and Philosophy* 20:697–713.

Porter, T. 2004. *Karl Pearson*. Princeton: Princeton University Press.

Power, Tom, Robert Stewart, Marie-Laure Ancelin, Isabelle Jaussent, Alain Malafosse, and Karen Ritchie. 2010. 5-HTTLPR Genotype, Stressful Life Events and Late-Life Depression: No Evidence of Interaction in a French Population. *Neurobiology of Aging* 31:886–887.

Preece, D. A. 1990. R.A. Fisher and Experimental Design: A Review. *Biometrics* 46:925–935.

Provine, William B. 1992. The R.A. Fisher-Sewall Wright Controversy. In *The Founders of Evolutionary Genetics: A Centenary Reappraisal*, ed. Sahotra Sarkar, 201–229. Dordrecht: Kluwer.

Provine, William B. 2001. *The Origins of Theoretical Population Genetics*. Chicago: University of Chicago Press.

Rader, Karen. 2014. *Making Mice: Standardizing Animals for American Biomedical Research, 1900–1955*. Princeton, NJ: Princeton University Press.

Raisman, Rita, Michael Briley, and Salomon Z. Langer. 1979. Specific Tricyclic Antidepressant Binding Sites in Rat Brain. *Nature* 281:148–150.

Ramamoorthy, Sammanda, Andrea L. Bauman, Kim R. Moore, Hong Han, Teresa Yang-Feng, Albert S. Chang, Vadivel Ganapathy, et al. 1993. Antidepressant- and Cocaine-Sensitive Human Serotonin Transporter: Molecular Cloning, Expression, and Chromosomal Localization. *Proceedings of the National Academy of Sciences of the United States of America* 90:2542–2546.

Ramasubbu, Rajamannar, Rose Tobias, Alastair M. Buchan, and N. Torben Bech-Hansen. 2006. Serotonin Transporter Gene Promoter Region Polymorphism Associated with Poststroke Major Depression. *Journal of Neuropsychiatry and Clinical Neurosciences* 18:96–99.

Rappaport, Maurice M., Arda Alden Green, and Irvine H. Page. 1948. Serum Vasoconstrictor (Serotonin): IV. Isolation and Characterization. *Journal of Biological Chemistry* 176:1243–1251.

Renwick, Chris. 2011. From Political Economy to Sociology: Francis Galton and the Social-Scientific Origins of Eugenics. *British Journal for the History of Science* 44:343–369.

Renwick, Chris. 2013. Completing the Circle of the Social Sciences? William Beveridge and Social Biology at London School of Economics during the 1930s. *Philosophy of the Social Sciences*. Published online before print 25 March 2013.

Retz, Wolfgang, Christine M. Freitag, Petra Retz-Junginger, Denise Wenzler, Marc Schneider, Christian Kissling, Johannes Thome, et al. 2008. A Functional Serotonin Transporter Promoter Gene Polymorphism Increases ADHD Symptoms in Delinquents: Interaction with Adverse Childhood Maltreatment. *Psychiatry Research* 158:123–131.

Richardson, Robert C. 1984. Biology and Ideology: The Interpretation of Science and Values. *Philosophy of Science* 51:396–420.

Risch, Neil, Richard Herrell, Thomas Lehner, Kung-Yee Liang, Lindon Eaves, Josephine Hoh, Andrea Griem, et al. 2009. Interaction between the Serotonin Transporter Gene (*5-HTTLPR*), Stressful Life Events, and Risk of Depression. *Journal of the American Medical Association* 301:2462–2471.

Risch, Neil, and Kathleen Merikangas. 1996. The Future of Genetic Studies of Complex Human Diseases. *Science* 273:1516–1517.

Robert, Jason Scott. 2004. *Embryology, Epigenesis, and Evolution: Taking Development Seriously*. Cambridge: Cambridge University Press.

Roll-Hansen, Nils. 1980. The Controversy between Biometricians and Mendelians: A Test Case for the Sociology of Scientific Knowledge. *Social Sciences Information–Information Sur les Sciences Sociales* 19:501–517.

Rosen, Christine. 2004. *Preaching Eugenics: Religious Leaders and the American Eugenics Movement*. Oxford: Oxford University Press.

Rothman, Kenneth J. 1978. Occam's Razor Pares the Choice among Statistical Models. *American Journal of Epidemiology* 108:347–349.

Rothman, Kenneth J., Sander Greenland, and Alexander M. Walker. 1980. Concepts of Interaction. *American Journal of Epidemiology* 112:467–470.

Royce, Kenneth. 1970. *The XYY Man*. London: Hodder & Stoughton.

Ruse, Michael. 1996. *Monad to Man: The Concept of Progress in Evolutionary Biology*. Cambridge, MA: Harvard University Press.

Rutter, Michael. 2008. Biological Implications of Gene-Environment Interaction. *Journal of Abnormal Child Psychology* 36:969–975.

Rutter, Michael. 2010. Gene-Environment Interplay. *Depression and Anxiety* 27:1–4.

Rutter, Michael, Henri Giller, and Ann Hagell. 1998. *Antisocial Behavior by Young People*. Cambridge: Cambridge University Press.

Rutter, Michael, Terrie E. Moffitt, and Avshalom Caspi. 2006. Gene-Environment Interplay and Psychopathology: Multiple Varieties but Real Effects. *Journal of Child Psychology and Psychiatry and Allied Disciplines* 47:226–261.

Rutter, Michael, Anita Thapar, and Andrew Pickles. 2009. Gene-Environment Interactions: Biologically Valid Pathway or Artifact? *Archives of General Psychiatry* 66:1287–1289.

Salmon, Wesley C. 1984. *Scientific Explanation and the Causal Structure of the World*. Princeton: Princeton University Press.

Saracci, Rodolfo. 1980. Interaction and Synergism. *American Journal of Epidemiology* 112:465–466.

Sarkar, Sahotra. 1996. Lancelot Hogben, 1895–1975. *Genetics* 142:655–660.

Sarkar, Sahotra. 1998. *Genetics and Reductionism*. Cambridge: Cambridge University Press.

Sarkar, Sahotra. 1999. From the *Reaktionsnorm* to the Adaptive Norm: The Norm of Reaction, 1909–1960. *Biology and Philosophy* 14:235–252.

Savulescu, Julian. 2001a. Procreative Beneficence: Why We Should Select the Best Children. *Bioethics* 15:413–426.

Savulescu, Julian. 2001b. Why Genetic Testing for Genes for Criminality Is Morally Required. *Princeton Journal of Bioethics* 4:79–97.

Savulescu, Julian, Melanie Hemsley, Ainsley Newson, and Bennett Foddy. 2006. Behavioural Genetics: Why Eugenic Selection is Preferable to Enhancement. *Journal of Applied Philosophy* 23:157–171.

Schaffner, Kenneth F. 1993. *Discovery and Explanation in Biology and Medicine*. Chicago: University of Chicago Press.

Schaffner, Kenneth F. 2001. Extrapolation from Animal Models: Social Life, Sex, and Super Models. In *Theory and Method in the Neurosciences*, ed. Peter K. Machamer, Rick Grush, and Peter McLaughlin, 200–230. Pittsburgh, PA: University of Pittsburgh Press.

Schaffner, Kenneth F. Forthcoming. Reduction: The Cheshire Cat Problem and a Return to Roots. *Synthese* 151:377–402.

Schaffner, Kenneth F. 2006. *Behaving: What's Genetic, and What's Not, and Why Should We Care?* Oxford: Oxford University Press.

Schaub, Bianca, Roger Lauener, and Erika von Mutius. 2006. The Many Faces of the Hygiene Hypothesis. *Journal of Allergy and Clinical Immunology* 117:969–977.

Scheid, J. M., C. B. Holzman, N. Jones, K. H. Friderici, K. A. Nummy, L. L. Symonds, A. Sikorskii, et al. 2007. Depressive Symptoms in Mid-Pregnancy, Lifetime Stressors and the 5-HTTLPR Genotype. *Genes, Brain and Behavior* 6:453–464.

Schmidt, Louis A., Nathan A. Fox, Kenneth H. Rubin, Stella Hu, and Dean H. Hamer. 2002. Molecular Genetics of Shyness and Aggression in Preschoolers. *Personality and Individual Differences* 33:227–238.

Schuster, Stephan C., Webb Miller, Aakrosh Ratan, Lynn P. Tomsho, Belinda Giardine, Lindsay R. Kasson, Robert S. Harris, et al. 2010. Complete Khoisan and Bantu Genomes from Southern Africa. *Nature* 463:943–947.

Seife, Charles. 2003. The Winner: Illuminating the Dark Universe. *Science* 302: 2038–2039.

Sen, Srijan, Margit Burmeister, and Debashis Ghosh. 2004. Meta-Analysis of the Association between a Serotonin Transporter Promoter Polymorphism (5-HTTLPR) and Anxiety-Related Personality Traits. *American Journal of Medical Genetics Part B: Neuropsychiatric Genetics* 127B:85–89.

Sen, Srijan, Henry R. Kranzler, John H. Krystal, Heather Speller, Grace Chan, Joel Gelernter, and Constance Guille. 2010. A Prospective Cohort Study Investigating Factors Associated with Depression during Medical Internship. *Archives of General Psychiatry* 67:557–565.

Sermon, K., V. Goossens, S. Seneca, W. Lissens, A. De Vos, M. Vandervorst, A. Van Steirteghem, et al. 1998. Preimplantation Diagnosis for Huntington's Disease (HD): Clinical Application and Analysis of the HD Expansion in Affected Embryos. *Prenatal Diagnosis* 18:1427–1436.

Sermon, Karen, Martine de Rijcke, Willy Lissens, Anick de Vos, Peter Platteau, Maryse Bonduelle, Paul Devroey, et al. 2002. Preimplantation Genetic Diagnosis for Huntington's Disease with Exclusion Testing. *European Journal of Human Genetics* 10:591–598.

Sesardic, Neven. 2005. *Making Sense of Heritability*. Cambridge: Cambridge University Press.

Shaw, E., and D. W. Woolley. 1956. Some Serotoninlike Activities of Lysergic Acid Diethylamide. *Science* 124:121–122.

Shimizu, Eiji, Kenji Hashimoto, and Masaomi Iyo. 2004. Ethnic Difference of the BDNF 196 G/A (val66met) Polymorphism Frequencies: The Possibility to Explain Ethnic Mental Traits. *American Journal of Medical Genetics Part B: Neuropsychiatric Genetics* 126B:122–123.

Shockley, William. 1972. The Apple-of-God's-Eye Obsession. *Humanist* 32:16–17.

Simpson, George Gaylord, Anne Roe, and Richard C. Lewontin. 1960. *Quantitative Zoology. Rev. Ed.* New York: Harcourt, Brace and Company.

Sjöberg, Rickard, Kent W. Nilsson, Niklas Nordquist, John Öhrvik, Jerzy Leppert, Leif Lindström, and Lars Oreland. 2006. Development of Depression: Sex and the Interaction between Environment and a Promoter Polymorphism of the Serotonin Transporter Gene. *International Journal of Neuropsychopharmacology* 9:443–449.

Sjoerdsma, Albert, and Michael G. Palfreyman. 1990. History of Serotonin and Serotonin Disorders. *Annals of the New York Academy of Sciences* 600:1–8.

Skipper, Robert A., Jr. 2002. The Persistence of the R.A. Fisher-Sewall Wright Controversy. *Biology and Philosophy* 17:341–367.

Skipper, Robert A., Jr. 2009. Revisiting the Fisher-Wright Controversy. *Transactions of the American Philosophical Society* 99:299–322.

Skipper, Robert A., and Roberta L. Millstein. 2005. Thinking about Evolutionary Mechanisms: Natural Selection. *Studies in the History and Philosophy of Biological and Biomedical Sciences* 36:327–347.

Slavishak, Edward. 2009. From Nation to Family: Two Careers in the Recasting of Eugenics. *Journal of Family History* 34: 89–115.

Smocovitis, Vassiliki (Betty). 2012. Humanizing Evolution: Anthropology, the Evolutionary Synthesis, and the Prehistory of Biological Anthropology, 1927–1962. *Current Anthropology* 53:S108–S125.

Sober, Elliott. 1980. Evolution, Population Thinking, and Essentialism. *Philosophy of Science* 47:350–383.

Solovieva, Svetlana, Jaana Lohiniva, Päivi Leino-Arjas, Raili Raininko, Katariina Luoma, Leena Ala-Kokko, and Hilkka Riihimäki. 2002. COL9A3 Gene Polymorphism and Obesity in Intervertebral Disc Degeneration of the Lumbar Spine: Evidence of Gene-Environment Interaction. *Spine* 27:2691–2696.

Soloway, Richard A. 1990. *Demography and Degeneration: Eugenics and the Declining Birthrate in Twentieth-Century Britain.* Chapel Hill: University of North Carolina Press.

Steel, Daniel P. 2008. *Across the Boundaries: Extrapolation in Biology and Social Science.* Oxford: Oxford University Press.

Stegenga, Jacob. 2011. Is Meta-Analysis the Platinum Standard of Evidence? *Studies in the History and Philosophy of Biological and Biomedical Sciences* 42:497–507.

Stein, Murray B., Nicholas J. Schork, and Joel Gelernter. 2008. Gene-by-Environment (Serotonin Transporter and Childhood Maltreatment) Interaction for Anxiety Sensitivity, an Intermediate Phenotype for Anxiety Disorders. *Neuropsychopharmacology* 33:312–319.

Stepan, Nancy Leys. 1991. *The Hour of Eugenics: Race, Gender, and Nation in Latin America*. Ithaca, NY: Cornell University Press.

Stochholm, Kirstine, Anders Bojesen, Anne Skakkebæk Jensen, Svend Juul, and Claus Højbjerg Gravholt. 2012. Criminality in Men with Klinefelter's Syndrome and XYY Syndrome: A Cohort Study. *BMJ Open* 2:e000650.

Stone, Robert D. 2003. The Cloudy Crystal Ball: Genetics, Child Abuse, and the Perils of Predicting Behavior. *Vanderbilt Law Review* 56:1557–1588.

Stotz, Karola. 2006. Molecular Epigenesis: Distributed Specificity as a Break in the Central Dogma. *History and Philosophy of the Life Sciences* 28:527–544.

Stotz, Karola. 2008. The Ingredients for a Postgenomic Synthesis of Nature and Nurture. *Philosophical Psychology* 21:359–381.

Stotz, Karola. 2012. Murder on the Development Express: Who Killed Nature/Nurture? A Review of Evelyn Fox Keller's *The Mirage of a Space between Nature and Nurture*, Duke University Press, 2010. *Biology and Philosophy* 27:919–929.

Strachan, David P. 1989. Hay Fever, Hygiene, and Household Size. *British Medical Journal* 299:1259–1260.

Surtees, Paul G., Nicholas W. J. Wainwright, Saffron A. G. Willis-Owen, Robert Luben, Nicholas E. Day, and Jonathan Flint. 2006. Social Adversity, the Serotonin Transporter (5-HTTLPR) Polymorphism and Major Depressive Disorder. *Biological Psychiatry* 59:224–229.

Suzuki, Yoichi, Satoshi Hattori, Yoichi Mashimo, Makiko Funamizu, Yoichi Kohno, Yoshitaka Okamoto, Akira Hata, et al. 2009. *CD14* and *IL4R* Gene Polymorphisms Modify the Effect of Day Care Attendance on Serum IgE Levels. *Journal of Allergy and Clinical Immunology* 123:1408–1411.

Swaminathan, Nikhil. 2007. "Got Smarts? Mother's Milk May Pump Up Baby's IQ." *Scientific American* (7 November). Available at http://www.scientificamerican.com/article.cfm?id=got-smarts-mothers-milk-m.

Tabery, James. 2004a. The "Evolutionary Synthesis" of George Udny Yule. *Journal of the History of Biology* 37:73–101.

Tabery, James. 2004b. Synthesizing Activities and Interactions in the Concept of a Mechanism. *Philosophy of Science* 71: 1–15.

Tabery, James. 2006. Looking Back on Lancelot's Laughter. *Mendel Newsletter* 15:10–17.

Tabery, James. 2007a. Biometric and Developmental Gene–Environment Interactions: Looking Back, Moving Forward. *Development and Psychopathology* 19: 961–976.

Tabery, James. 2007b. *Causation in the Nature-Nurture Debate: The Case of Genotype-Environment Interaction*. PhD Dissertation. Pittsburgh: University of Pittsburgh.

Tabery, James. 2008. R. A. Fisher, Lancelot Hogben, and the Origin(*s*) of Genotype-Environment Interaction. *Journal of the History of Biology* 41:717–761.

Tabery, James. 2009a. Difference Mechanisms: Explaining Variation with Mechanisms. *Biology and Philosophy* 24:645–664.

Tabery, James. 2009b. Making Sense of the Nature-Nurture Debate: Review of Neven Sesardic (2005), *Making Sense of Heritability*. *Biology and Philosophy* 24:711–723.

Tabery, James. 2009c. From a Genetic Predisposition to an *Interactive Predisposition*: Rethinking the Ethical Implications of Screening for Gene-Environment Interactions. *Journal of Medicine and Philosophy* 34:27–48.

Tabery, James. 2009d. Interactive Predispositions. *Philosophy of Science* 76: 876–888.

Tabery, James. 2011. Commentary: Hogben vs. the Tyranny of Averages. *International Journal of Epidemiology* 40:1454–1458.

Tabery, James, and Paul E. Griffiths. 2010. Historical and Philosophical Perspectives on Behavior Genetics and Developmental Science. In *Handbook of Developmental Science, Behavior, and Genetics*, ed. Kathryn E. Hood, Carolyn Tucker Halpern, Gary Greenberg, and Richard M. Lerner, 41–60. Malden, MA: Blackwell Publishing.

Tal, Omri. 2009. From Heritability to Probability. *Biology and Philosophy* 24: 81–105.

Tal, Omri. 2012. The Impact of Gene-Environment Interaction and Correlation on the Interpretation of Heritability. *Acta Biotheoretica* 60:225–237.

Taylor, Alan, and Julia Kim-Cohen. 2007. Meta-Analysis of Gene-Environment Interactions in Developmental Psychopathology. *Development and Psychopathology* 19:1029–1037.

Taylor, Norman B. 1931. The Relation of Temperature to the Heart Rate of the South African Frog (*Xenopus Dactylethra*). *Journal of Physiology* 71:156–168.

Taylor, Peter. 2006. Heritability and Heterogeneity: On the Limited Relevance of Heritability in Investigating Genetic and Environmental Factors. *Biological Theory* 1:150–164.

Taylor, Peter. 2010. Three Puzzles and Eight Gaps: What Heritability Studies and Critical Commentaries Have Not Paid Enough Attention To. *Biology and Philosophy* 25:1–31.

Taylor, Shelley E., Baldwin M. Way, William T. Welch, Clayton J. Hilmert, Barbara J. Lehman, and Naomi I. Eisenberger. 2006. Early Family Environment, Current

Adversity, the Serotonin Transporter Promoter Polymorphism, and Depressive Symptomatology. *Biological Psychiatry* 60:671–676.

Teigen, Karl H. 1984. A Note on the Origin of the Term "Nature and Nurture": Not Shakespeare and Galton, but Mulcaster. *Journal of the History of the Behavioral Sciences* 20:363–364.

Tercyak, Kenneth P., Sharon Hensley Alford, Karen M. Emmons, Isaac M. Lipkus, Benjamin S. Wilfond, and Colleen M. McBride. 2011. Parents' Attitudes Toward Pediatric Genetic Testing for Common Disease Risk. *Pediatrics* 127:e1288.

Thompson, E. A. 1990. R. A. Fisher's Contributions to Genetical Statistics. *Biometrics* 46:905–914.

Tienari, Pekka, Lyman C. Wynne, Juha Moring, Ilpo Lahti, Mikko Naarala, Anneli Sorri, Karl-Erik Wahlberg, et al. 1994. The Finnish Adoptive Family Study of Schizophrenia: Implications for Family Research. *British Journal of Psychiatry* 164 (Suppl. 23):20–26.

Tozzi, Federica, Alexander Teumer, Marcus Munafò, Rajesh Rawal, Gbenga Kazeem, Marcel Gerbaulet, Wendy McArdle, et al. 2013. A Genomewide Association Study of Smoking Relapse in Four European Population-Based Samples. *Psychiatric Genetics* 23:143–152.

Tucker-Drob, Elliot M., Mijke Rhemtulla, K. Paige Harden, Eric Turkheimer, and David Fask. 2011. Emergence of a Gene×Socioeconomic Status Interaction on Infant Mental Ability between 10 Months and 2 Years. *Psychological Science* 22:125–133.

Turkheimer, Eric. 1998. Heritability and Biological Explanation. *Psychological Review* 105:782–791.

Turkheimer, Eric. 2000. Three Laws of Behavior Genetics and What They Mean. *Current Directions in Psychological Science* 9:160–164.

Turkheimer, Eric. 2008. A Better Way to Use Twins for Developmental Research. *LIFE Newsletter* 2:2–5.

Turkheimer, Eric. 2011. Commentary: Variation and Causation in the Environment and Genome. *International Journal of Epidemiology* 40:598–601.

Turkheimer, Eric. 2012. Genome Wide Association Studies of Behavior are Social Science. In *Philosophy of Behavioral Biology: Boston Studies in the Philosophy of Science*, vol. 282, ed. Katie S. Plaisance and Thomas A. C. Reydon. New York: Springer.

Turkheimer, Eric, and Irving I. Gottesman. 1991. Is *H2* = 0 a Null Hypothesis Anymore? *Behavioral and Brain Sciences* 14:410–411.

Turkheimer, Eric, Andreana Haley, Mary Waldron, Brian D'Onofrio, and Irving I. Gottesman. 2003. Socioeconomic Status Modifies Heritability of IQ in Young Children. *Psychological Science* 14:623–628.

Turkheimer, Eric, and Erin E. Horn. 2014. Interactions between Socioeconomic Status and Components of Variation in Cognitive Ability. In *Behavior Genetics of Cognition Across the Lifespan*, ed. D. Finkel and C. A. Reynolds. New York: Springer.

Twarog, Betty Mack, and Irvine H. Page. 1953. Serotonin Content of Some Mammalian Tissues and Urine and a Method for Its Determination. *American Journal of Physiology* 175:157–161.

Tyler, William J., Mariana Alonso, Clive R. Bramham, and Lucas D. Pozzo-Miller. 2002. From Acquisition to Consolidation: On the Role of Brain-Derived Neurotrophic Factor Signaling in Hippocampal-Dependent Learning. *Learning & Memory (Cold Spring Harbor, NY)* 9:224–237.

Uher, R., and P. McGuffin. 2008. The Moderation by the Serotonin Transporter Gene of Environmental Adversity in the Aetiology of Mental Illness: Review and Methodological Analysis. *Molecular Psychiatry* 13:131–146.

University of Michigan Health System. 2011. "Resurrecting the So-Called 'Depression Gene': New Evidence That Our Genes Play a Role in Our Response to Adversity." *ScienceDaily* (3 January). Available at http://www.sciencedaily.com/releases/2011/01/110103161105.htm.

U.S. Department of Health and Human Services, Children's Bureau. 2012. *Child Maltreatment* 2011. available online http://www.acf.hhs.gov/programs/cb/research-data-technology/statistics-research/child-maltreatment.

Vandenberg, Steven G., ed. 1966. *Progress in Human Behavior Genetics: Recent Reports on Genetic Syndromes, Twin Studies, and Statistical Analyses*. Baltimore, MD: The Johns Hopkins Press.

Van IJzendoorn, Marinus H. 1995. Adult Attachment Representations, Parental Responsiveness, and Infant Attachment: A Meta-Analysis on the Predictive Validity of the Adult Attachment Interview. *Psychological Bulletin* 117:387–403.

Veletza, Stavroula, Maria Samakouri, George Emmanouil, Gregory Trypsianis, Niki Kourmouli, and Miltiadis Livaditis. 2009. Psychological Vulnerability Differences in Students—Carriers or Not of the Serotonin Transporter Promoter Allele S: Effect of Adverse Experiences. *Synapse (New York, NY)* 63:193–200.

Vercelli, Donata. 2010. Gene-Environment Interactions in Asthma and Allergy: The End of the Beginning? *Current Opinion in Allergy and Clinical Immunology* 10:145–148.

Vinovskis, Maris. 2005. *The Birth of Head Start: Preschool Education Policies in the Kennedy and Johnson Administrations*. Chicago: University of Chicago Press.

Visscher, Peter M., Matthew A. Brown, Mark I. McCarthy, and Jian Yang. 2012. Five Years of GWAS Discovery. *American Journal of Human Genetics* 90:7–24.

Von Mutius, Erika. 2007. Allergies, Infections and the Hygiene Hypothesis—The Epidemiological Evidence. *Immunobiology* 212:433–439.

Von Mutius, Erika. 2009. Gene-Environment Interactions in Asthma. *Journal of Allergy and Clinical Immunology* 123:3–11.

Vreeke, G.-J. 2000. Nature, Nurture, and the Future of the Analysis of Variance. *Human Development* 43: 32–45.

Waddington, Conrad H. 1957. *The Strategy of the Genes*. London: Allen and Unwin.

Wagman, Paul. 1973. "The Brains Do Battle in I.Q. Controversy." *The Boston Phoenix* (November 6): 18, 28.

Wahlberg, Karl-Erik, Lyman C. Wynne, Hannu Oja, Pirjo Keskitalo, Liisa Pykäläinen, Ilpo Lahti, Juha Moring, et al. 1997. Gene-Environment Interaction in Vulnerability to Schizophrenia: Findings from the Finnish Adoptive Family Study of Schizophrenia. *American Journal of Psychiatry* 154:355–362.

Wahlsten, Douglas. 1990. Insensitivity of the Analysis of Variance to Heredity-Environment Interaction. *Behavioral and Brain Sciences* 13:109–161.

Wahlsten, Douglas. 2000. Analysis of Variance in the Service of Interactionism. *Human Development* 43:46–50.

Walter, S. D., and T. R. Holford. 1978. Additive, Multiplicative, and Other Models for Disease Risk. *American Journal of Epidemiology* 108:341–346.

Wang, Shirley S. 2009. "Serotonin Gene's Link with Depression Not Valid, Says Study." *The Wall Street Journal* (16 June). Available at http://blogs.wsj.com/health/2009/06/16/serotonin-genes-link-with-depression-not-valid-says-study/.

Wasserman, David. 2004. Is There Value in Identifying Individual Genetic Predispositions to Violence? *Journal of Law, Medicine & Ethics* 32:24–33.

Waters, C. Kenneth. 2007. Causes That Make a Difference. *Journal of Philosophy* CIV:551–579.

Weber, Marcel. 2005. *Philosophy of Experimental Biology*. Cambridge: Cambridge University Press.

Weiss, Sheila Faith. 2010. *The Nazi Symbiosis: Human Genetics and Politics in the Third Reich*. Chicago: University of Chicago Press.

Wellman, C. L., A. Izquierdo, J. E. Garrett, K. P. Martin, J. Carroll, R. Millstein, K.-P. Lesch, et al. 2007. Impaired Stress-Coping and Fear Extinction and Abnormal Corticolimbic Morphology in Serotonin Transporter Knock-Out Mice. *Journal of Neuroscience* 17:684–691.

Wells, G. P. n.d. "Father and Son." Lancelot Hogben Papers (A.38), University of Birmingham.

Wells, G. P. 1978. Lancelot Thomas Hogben. *Biographical Memoirs of Fellows of the Royal Society of London* 24:183–221.

Werskey, Gary. 1978. *The Visible College: The Collective Biography of British Scientific Socialists of the 1930s*. New York: Holt, Rinehart and Winston.

Wertz, Dorothy C., Joanna H. Fanos, and Philip R. Reilly. 1994. Genetic Testing for Children and Adolescents: Who Decides? *Journal of the American Medical Association* 272:875–881.

Wheeler, David A., Maithreyan Srinivasan, Michael Egholm, Yufeng Shen, Lei Chen, Amy McGuire, Wen He, et al. 2008. The Complete Genome of an Individual by Massively Parallel DNA Sequencing. *Nature* 452:872–876.

Whitaker-Azmitia, Patricia Mack. 1999. The Discovery of Serotonin and Its Role in Neuroscience. *Neuropsychopharmacology* 21:2S–8S.

Whitney, Glayde. 1990. A Contextual History of Behavior Genetics. In *Developmental Behavior Genetics: Neural, Biometrical, and Evolutionary Approaches*, ed. Martin E. Hahn, John K. Hewitt, Norman D. Henderson, and Robert H. Benno. New York: Oxford University Press.

Wilhelm, Kay, Philip B. Mitchell, Heather Niven, Adam Finch, Lucinda Wedgwood, Anna Scimone, Ian P. Blair, et al. 2006. Life Events, First Depression Onset and the Serotonin Transporter Gene. *British Journal of Psychiatry* 188: 210–215.

Wilson, Edward O. 1975. *Sociobiology: The New Synthesis*. Cambridge, MA: Harvard University Press.

Wilson, Jim. 2002. Criminal Genes. *Popular Mechanics* 179:46–48.

Wimsatt, William. 1976. Reductive Explanation: A Functional Account. In *Conceptual Issues in Evolutionary Biology*, ed. Elliott Sober, 369–385. Cambridge, MA: MIT Press.

Winton, Frank. 1927. A Contrast between the Actions of Red and White Squills. *Journal of Pharmacology and Experimental Therapeutics* 31:137–144.

Woodward, James. 2003. *Making Things Happen: A Theory of Causal Explanation*. Oxford: Oxford University Press.

Woolley, D. W., and E. Shaw. 1954. A Biochemical and Pharmacological Suggestion about Certain Mental Disorders. *Proceedings of the National Academy of Sciences of the United States of America* 40:228–231.

Yaffe, K., M. Haan, A. Byers, C. Tangen, and L. Kuller. 2000. Estrogen Use, *APOE*, and Cognitive Decline: Evidence of Gene-Environment Interactions. *Neurology* 54:1949–1953.

Yamada, Yukio. 1960. Observations of Genotype-Environment Interaction in Productive Traits in Chickens. *Japanese Journal of Breeding* 10:23–28.

Yates, F. 1964. Sir Ronald Fisher and the Design of Experiments. *Biometrics* 20:307–321.

Yates, F., and K. Mather. 1963. Ronald Aylmer Fisher. *Biographical Memoirs of Fellows of the Royal Society of London* 9:91–120.

Yule, George Udny. 1902. Mendel's Laws and Their Probable Relations to Intra-Racial Heredity. *New Phytologist* 1:193–207, 222–238.

Zalsman, Gil, Yung-Yu Huang, Maria A. Oquendo, Ainsley K. Burke, Xian-Zhang Hu, David A. Brent, Steven P. Ellis, et al. 2006. Association of a Triallelic Serotonin Transporter Gene Promoter Region (5-HTTLPR) Polymorphism with Stressful Life Events and Severity of Depression. *American Journal of Psychiatry* 163:1588–1593.

Zammit, Stanley, Michael J. Owen, and Glyn Lewis. 2010. Misconceptions about Gene-Environment Interactions in Psychiatry. *Evidence-Based Mental Health* 13:65–68.

Zhang, Kerang, Qi Xu, Yong Xu, Hong Yang, Jinxiu Luo, Yan Sun, Ning Sun, et al. 2009. The Combined Effects of the 5-HTTLPR and 5-HTR1A Genes Modulates the Relationship between Negative Life Events and Major Depressive Disorder in a Chinese Population. *Journal of Affective Disorders* 114:224–231.

Zhou, Wei, Geoffrey Liu, David P. Miller, Sally W. Thurston, Li Lian Xu, John C. Wain, Thomas J. Lynch, et al. 2002. Gene-Environment Interaction for the *ERCC2* Polymorphism and Cumulative Cigarette Smoking Exposure in Lung Cancer. *Cancer Research* 62:1377–1381.

Zigler, Edward, and Sally J. Styfco. 2010. *The Hidden History of Head Start*. Oxford: Oxford University Press.

Index

Printed in the United States
by Baker & Taylor Publisher Services